SHAKESPEARE & JUNG
THE GOD IN TIME

Meditations on Time, God
& our Value Creating Universe

SHAKESPEARE & JUNG
THE GOD IN TIME

Meditations on Time, God & our Value Creating Universe

With: A Philosophical and Literary Glossary

By James P. Driscoll, Ph.D.

ACADEMICA PRESS
LONDON-WASHINGTON

Library of Congress Cataloging-in-Publication Data

Names: Driscoll, James P., 1943- author.
Title: Shakespeare & Jung - the God in time : meditations on time, God, & our
value creating universe : with a philosophical and literary glossary / by James P. Driscoll, Ph.D.
Other titles: Shakespeare and Jung - the God in time
Description: Washington : Academica Press, 2019. | Includes bibliographical
references and index.
Identifiers: LCCN 2018056075 | ISBN 9781680534818 (hardcover : alk. paper)
Subjects: LCSH: Time. | God. | Shakespeare, William, 1564-1616. | Jung, C. G.
(Carl Gustav), 1875-1961. | Whitehead, Alfred North, 1861-1947.
Classification: LCC BD638 .D75 2019 | DDC 211--dc23
LC record available at https://lccn.loc.gov/2018056075

Thanks to the University of Kentucky Press for permission to use sections of the Glossary from my 1993 book, *The Unfolding God of Jung and Milton.*

CONTENTS

PREFACE
(why read this book?)

"What is new or different in your book? What reader interests does it speak to?" An author may be the best and worst person to answer these questions. Any author should have thoroughly studied the literature on his subjects. He should know how he differs from other work in the area, complements it, enhances and advances it, rebuts, or corrects it. Still, authors are usually the least objective critics of their own work. Asking an author to evaluate the appeal of her book is like asking a mother to evaluate the appeal of her newborn baby.

Hazards notwithstanding, I shall endeavor to briefly answer these questions. First, *The God in Time* addresses multiple reader interests exploring the intersections of psychology, philosophy, religion, and literature. It presents novel perspectives on Shakespeare, C.G Jung, and A.N. Whitehead and on their shared philosophical, religious, and psychological concerns. Chief among these are the nature of God and time, their relationships, and their broad implications.

What can the reader learn here that he will not find elsewhere? Some will argue that there are no really new ideas on any major philosophical, religious, literary, or psychological topic. Even if that is in a general sense true, there are always fresh approaches to and combinations of known ideas that can strengthen, expand, and deepen our understanding. One can gain new insight into Jung's, Whitehead's, or Shakespeare's ideas on God and time, by connecting, comparing, and contrasting them or by trying to view the ideas of one thinker through the eyes of another.

Let me outline some principle ideas and themes whose treatment herein yields innovative perspectives. This will be only a bare sketch: detailed development is the task of the book itself.

1.Time's irreversibility and continuous creation of novelty makes Time the source of value and ground of meaning, an essential feature of actual God and proof of God's existence. Time interconnects and differentiates all and thereby makes reality relational allowing for thought and logic.

2. The term "God" has five key levels of meaning. The most basic is the God image within the psyche of each person, then come the tribal and canonical conceptions of deity. Third is the civilizational Godhead, e.g., Allah, Zeus, or the Christian Godhead identified with the Trinity but energizing by four different archetypes according to Jung. Fourth is Nature itself, Gaia, the ruling deity of the earthly biosphere. Fifth and highest is Actual God, the controlling forces and principles of the universe. Failure to heed these distinctions is the source of much confusion.

3. The archetypes energizing the Judeo-Christian Godhead are the Father, the bifurcate hostile brothers (The Son versus Prometheus-Lucifer), and the wiseperson or *anima* figure. There are three paradigmatic

works in the revelation of the Western Godhead: The Book of Job, The Gospels, and Shakespeare's *King Lear*. In our time Jung's *Answer to Job* attempted to wake us from our dogmatic slumbers on the Godhead which, he argued, must be a Quaternity, not a Trinity. But Shakespeare penetrated furthest into the dynamics of the Godhead giving us our most powerful and comprehensive revelation of its dominant archetypes.

4. Time holds singular importance in Shakespeare, indeed time is the de facto God of Shakespeare's worlds. Shakespeare above all, but far from alone, illumines great philosophical and religious questions through his artist's vision. The entire human artistic canon, topped by Shakespeare and including great painters and musicians, is our touchstone for the truths of the human condition and our chief gate to wisdom. Whitehead and Shakespeare understood that the creation of beauty, value, meaning, and enjoyment is the supreme gift of The God in Time. Jung's work on synchronicity challenges us to deepen understanding of time and the psyche.

5. Entropy and coalescence are two fundamental opposing forces in time and the universe. Life is the supreme accomplishment of coalescence. Coalescence manifests the creativity of The God in Time. Coalescence parallels Whitehead's concept of concrescence and Teilhard de Chardin's complexification. At the level of personal psychology, Jung's individuation process exemplifies coalescence.

6. The archetypes of the Western Godhead energized and shaped the development of science and technology along with the ideals of progress and human freedom. The West developed differently and

advanced more rapidly because of the unique archetypal dynamics of our Godhead.

7. Basic positions of philosophers fall in a spectrum that corresponds to time's past, present, and future tenses. Stasis philosophers are past oriented, flux philosophers are future oriented, and process philosophers stress the present. A philosopher's stance on time is always pivotal to his other stances.

8. Actual God, The God in Time, makes, by the irreversibility of time, a universe where suffering, injustice, irreparable loss, tragedy and death are inevitable. Yet the creative nature of time renders value, beauty, enjoyment, and meaning not only possible, but evident to anyone. The human condition affords eight essential guidelines for optimizing the good and overcoming evils by learning to live in harmony with The God in Time.

There is much more in *Shakespeare & Jung--The God in Time* beyond these points, so if they arouse interest, this book should reward your curiosity and diligence.

INTRODUCTION

Thou'lt come no more,
Never, Never, Never, Never, Never! - King Lear V. iii

"At the heart of the nature of things there is always the dream of youth and the harvest of tragedy."- A.N. Whitehead.

Rejecting time's irreversibility for personal immortality in an eternity beyond time is a flight from the tragic realities of life. All things arise, grow, flourish, and perish in their own special places in irreversible time: irreversibility makes them unique and gives them value and meaning. Much of the history of philosophy and religion is a story of man's attempts to deny The God in Time. - The Author.

Does God exist? Every thoughtful person asks this question, and comes back to it. It's the biggest question of all, the beginning and end of many discussions on religion, philosophy, and the meaning of life, and of death. The existence of God should not be something mysterious or difficult to understand. It should be obvious to everyone, even though many do not recognize it. The god-image, Carl Jung maintains, exists in the psyche of every individual. A pantheistic God of the cosmos exists because Time exists. Time creates order, value, and meaning through temporal relationships everywhere all the time. That is what time does, and creation of order, value, and meaning is what God does. So the proof of the nature and existence of a cosmic God lies in the nature and regulatory function of time. Newton's famous pronouncement, *hypotheses non fingo*, tells us, though it did not sink in with Newton, that we need look no further than time itself for find the God of the universe.

People's resistance to believing that God operates in time and is Itself time is at bottom resistance to accepting time's fundamental irreversibility, which is to say time's absolute power, boundless creativity, and ultimate impersonality. The God in Time is the context for the more limited, more personal, culturally determined visions of deity prevalent in human societies, not an alternative to them. Amplifying these points and their ramifications will be one of the chief tasks undertaken in these pages.

The task proceeds through a series of meditations that view humanity's recurrent questions about and challenges with time from diverse perspectives, refining and enlarging them with each return. The progression of argument does not, therefore, move in a strictly linear fashion from the beginning to the end of the book. The meditations reflect on aspects of time and God and on different problems key thinkers, as well as ordinary people, encounter dealing with time and God. They are divided into numbered sections, each dealing with separate facets of their general topic. Again, the overall structure is more like a painting than rising in a stairway-like logical or mathematical progression. The method is to advance imaginative apprehension through multiple perspectives.

Thus, critical points are repeated in different places and shown in differing lights and contexts to broaden, deepen, and refine understanding of them and of the God in time.

The ideas of two preeminent thinkers, Carl Jung particularly, then Alfred North Whitehead and above all the supreme poet and dramatist of all ages, William Shakespeare, have influenced these meditations throughout. While Whitehead is cited somewhat less frequently, in part because of the difficulty of his ideas and terminology, his thought provides an essential metaphysical background and cosmological framework. My purpose is not as much to illumine Jung's, Whitehead's, or even Shakespeare's ideas on time as it is to use each to enhance and deepen the others, toward the objective of manifesting all their insights, and my own, to illumine time itself and, ultimately, unveil the God in time.

Although disadvantaged by being born before Shakespeare, Plato and Aristotle credited the philosophic import of serious poetry. Aristotle wrote: "Hence poetry is something more philosophic and of graver import than history, since its statements are of the nature rather of universals, whereas those of history are singulars.". Aristotle's insight on the value and use of literature was pursued by only a few of the philosophers who succeeded him. However, Sigmund Freud and Carl Jung both recognized it. Their applications were limited, understandably because philosophy is another field peripheral to their main endeavors. They left it to others who could more evenly divide their attentions between literature, philosophy, and psychology. As one of those, after many years of intensive study of both Jung and Shakespeare I reached the conclusion that we can often learn as much about Jung's concerns from Shakespeare as we can learn about Shakespeare from Jung. For example, the study of Shakespeare can illumine one of the subjects that most fascinated and troubled Jung, the nature and function of the archetypes powering the Judeo-Christian Godhead.

No other great writer is as preoccupied with time as Shakespeare. He has a broader range of insights on time than any other writer or philosopher. For the most part, these are brilliantly stated truisms rather than "original insights." But the most valuable form of originality is often just saying what many have always known better than any have said it before. Hamlet's "the readiness is all" may not at first seem profound, but it becomes so in the context of the play, Shakespeare's opus, and human experience. One needs to take all the Shakespearean comments on time, that is apprehend his entire vision of time, in order to appreciate its extraordinary scope and grasp its profundities. Few literary critics try to do this. They are especially reluctant in this postmodern age where many come not to praise Shakespeare but to bury him, either by treating his works as historical artifacts, like moldering tomes left in the stacks, or as cadavers to be autopsied, or deconstructed, which in their practice often means diminished and debunked.

Freud had the insight and imagination to grasp deep psychological meanings in two of Western literature's paradigmatic tragedies, *Oedipus Rex* and *Hamlet*. Though not technically a tragedy, *The Book of Job* stands with *Oedipus Rex* as one of four preeminent stories at the heart of the tragic vision of the West, the other two being the gospel narrative and *King Lear*.

In *Answer to Job,* Jung approached Job's story in relation to the gospel narrative, attempting to do for the Judeo-Christian Godhead what Freud attempted to do for human sexual psychology with *Oedipus Rex*.

Before my article on *King Lear* and my *Identity in Shakespearean Drama* of thirty some years ago, it appears to have occurred to few, if any, that *King Lear* dramatizes the archetypes of the Godhead that preoccupied Jung and other deep thinkers concerned with theodicy. My insight has so far not received wide attention; that is due, I suspect, mainly to three factors. First, academia today is compartmentalized so that psychologists, philosophers, and theologians seldom pay heed to either literary critics or to each other. Second, at the time my ideas were first published there was determined opposition to Jungian approaches from old school academics. Their fixed dogma held that meanings Shakespeare did not consciously intend could not be in his plays. For them meaning in literature is entirely a rational construct, the role of imagination is simply to shine it up: the backstory from the artist's psyche is ignored or denied. Third, the old school was displaced in the 80s by newcomers committed to revisionist ideological approaches driven by political agendas.

After a long absence from criticism, events impelled me to revisit my work on *King Lear* in relation to time and the Godhead. I had dealt with the issues around Godhead and theodicy in depth in my 1993 book, *The Unfolding God of Jung and Milton.* The current work differs from its predecessors in that there is a lot more philosophy, particularly Whitehead, and the focus is on time and God and their essential relationships.

My undergraduate major was philosophy, I switched later in graduate school to English. My tardiness, due to strong interest in the philosophical issues in this work, gave me a background in philosophy. Mythology and archetypal psychology became a lifelong avocation while writing my Shakespeare dissertation. Later I had several years of Jungian analysis. So much for my contemplative life, unlike most academic writers I also have had a three decade long active life as an advocate for HIV patients, FDA and healthcare reforms, and LGBT rights. My active life has broadened and deepened my experiences with and perspectives upon the perennial philosophic issues surrounding evil, suffering, death, and injustice. I have seen firsthand and often how individuals' existential struggles impact their quests for meaning and value in their lives.

Drawing on Shakespeare and Jung, along with Whitehead, to explore the nature of God and time is an approach unique to this work. The interdependence of God and Time in the generation of value and meaning, if not a concept entirely original to this work, is unusual in the extent of its treatment and variety of angles of approach. While a philosophical exploration of time must take account of Newton, Darwin, and Einstein, I found that quantum physics is not yet the open Sesame some seek. No scientific, or philosophic, writing on time equals Shakespeare's in depth, breadth, and relevance to timeless human concerns. Some may demur that I give short shrift to classic philosophers, such as Plato, Spinoza, and Kant. The brief treatment stems from my belief that the limits of their knowledge of psychology, and of the universe itself, restricts their perspectives on the questions I address. For example, the above three were largely incognizant of the vast realm of the unconscious, and lacked the conceptual tools honed by Freud, Jung, and others in the last century and a half.

Nor could they draw upon the intellectual capital bequeathed by modern physics, astronomy, geology, anthropology, and biology, especially evolution, the one idea that has most profoundly altered man's view of himself, his place in nature, and the timeframe of creation. To put it another way, God's revelations to man are limited by man's capacity to understand. All these modern advances have enhanced that capacity. Yet, except for Jung and Whitehead, among major intellectual figures of the last century there has been little effort to use them to deepen understanding of deity. Therefore, I wanted to move quickly beyond the limitations of previous ages to base my approach to God and time upon the whole of human experience to date, especially that from the arts, an area too often neglected.

A familiar short coming of contemporary academic philosophers and psychologists is their penchant for disputing with one another instead of learning from the arts and the great writers outside philosophy who address, from their individual perspectives, the basis in human experience of philosophical problems like the nature of God and time. Another common, if seldom recognized, deficiency is reluctance, or inability, to draw out the implications of their positions for contemporary social and political problems. They are wont to bemoan the spiritual and moral crisis of the West without venturing to address actual problems or specific solutions. To avoid such evasions, at the conclusion of *The God in Time* I review several contemporary dilemmas that are rooted in limited religious understanding. I then present my observations on the human condition itself that may be drawn from the treatment of God and time.

This work is organized as a collection of twenty-four meditations; most have subsections, an exception being the piece on Shakespeare which is a long essay. Several years ago I began writing notes, usually in the morning, recording my thoughts on time and God. Earlier in life I mused on writing about time, a subject that fascinated me. But the most appropriate point for such an undertaking seemed to be old age after one had experienced the major part of a human life. At age 69 I had a near fatal heart attack in an airplane on my way to Washington DC. A cardiologist on board took charge demanding that the pilot turn the plane around and land in Albuquerque which had an excellent trauma center next to the airport. Without his professional zeal on behalf of a patient he didn't know, this work might never have been undertaken. Just as Jung's thoughts turned more to God, the Godhead, and time after his first heart attack, so did mine.

The God in Time bears limited similarity to Marcus Aurelius *Meditations* in that my cogitations were written separately, and not initially intended to form a book. But the ideas herein are deeply influenced by two angry works whose attitudes are far from stoical: Shakespeare's *King Lear* and Jung's *Answer to Job. King Lear,* I believe, answers some of Jung's unanswered questions in *Answer to Job.* My goals are to reward the open minded with deeper understanding of The God in Time, and to illumine the archetypes powering the Judeo-Christian

Godhead, their relationships to time, and their staggering impact on Western civilization

.

MEDITATION I:
TIME ITSELF

1. Many religious thinkers seek meaning and value outside of or beyond time, as if dissatisfied with the meaning of human life within "the vail of tears" that is time. In the West they typically identify the source of meaning as a "timeless" Supreme Being who created time, yet resides outside time in eternity. They fail to grasp that a meaning beyond time is at best a redundancy because we all depend on and are defined by our lives or existence in our temporal context. Everything outside of it is speculation. They fail to consider that we are in eternity right now, but simply do not ordinarily recognize it. Eternity is not beyond time: in the present instant is eternity. Meaning involves conscious reflection upon, among other things, relationships, conflicts, and conflict resolution which exist only in time. Value involves feeling, feelings are ever fleeting, that is temporal. There are no meanings, thoughts, feelings or values in an eternity beyond time, Plato and the Hindu seers notwithstanding. Such an eternity is a conjecture defined by/against time, not a verifiable experience we can sense or measure; in short, eternity beyond time does not "exist." It functions like a zero in the number system where the numbers are time itself.

2. Augustine and subsequent Christian theologians, dreaming Plato's dream of eternal, unchanging forms, assumed that time and all its contents find their ultimate meaning, value, and purpose in reference to eternity. Heidegger, closer to the realities of the lives we live, thought we find meaning in our temporal existence, our historicity as he called it, and ultimately in the termination of our existences, death. But death is only one, though uniquely important, moment in the flow of time. The experience of death forces us to confront, and nurtures awareness of, our finite existence. Yet the meaning of a death is defined by the life that precedes it, and by that life's impact on the future.

Meaning is the character and quality of a sequence of events in time, shown in reflection upon their origins, intensity, relationships, consequences, but above all their uniqueness and singular place interwoven within the processes that constitute the web of time. Attempts to find meaning often endeavor to state or express the essential character of a moment or a discrete period of time. Velasquez's masterpiece *Las Meninas* shows how a great painter can capture the dense web of meaning epitomized in a moment. Strictly speaking, time itself has no meaning or value, it is the medium or matrix, the canvas of the painting, wherein meaning exists and value is created. Moreover, time is the dynamism of reality itself. Meaning, value, and time depend on dynamic change, that is creation, duration, conflict, and destruction. Time is the medium and driver of the dynamism of change and, therefore, of meaning and value.

3. "In the moment there is nothing," observed Alfred North Whitehead. True in the sense that nothing changes in a moment. However,

moments are theoretical constructs, they are the mind's attempt to impose order on the seamless flow, or continuum, of perception that is time. Perhaps equally true, "in the moment there is everything" since the past no longer exists, the future has not yet come into being, so we have only the moment. But a reflective moment, whether in man or beast, is shaped by a context formed from recollection of the immediate past and anticipation of the immediate future. Without the contexts provided by recollections and anticipations, the moment would implode upon itself like a black hole. Consciousness exists only in the moment, yet it cannot exist without recollecting a context in the past or anticipating a future context. Without recollection and anticipation, the present would be a meaningless blur of chaotic sensation, like a wild ride on a psychic roller coaster. The sole thing that does exist, the present moment, depends on things that no longer exist, the past, or that do not yet exist, the future. That is one of time's conumdrums, only one out of many. Time is the great, over-riding, immanent, imminent, and transcendent mystery replete with conundrums.

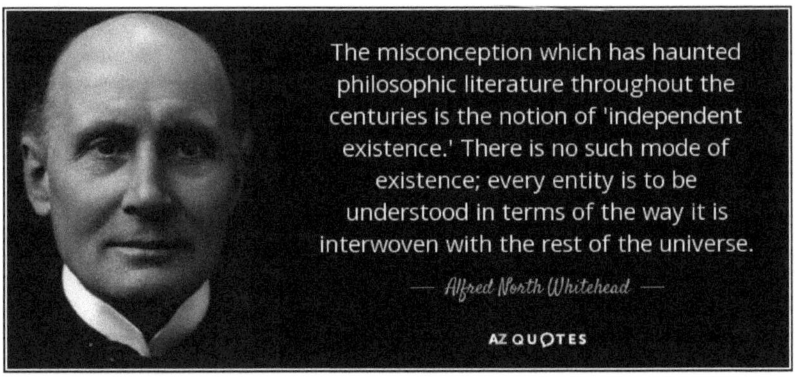

4. A dynamic universe with movement and spatial separation or non-identity between composite entities must necessarily be a temporal, process universe, or what A.N. Whitehead referred to as a *creative*

continua. Movement both creates and requires space and time. It is impossible for human minds to conceive a universe that has movement and separate entities but is without time. Movement among separate entities creates dynamic relationships which are essential to consciousness and to thought as well as perception and feeling. A relational universe must be a temporal universe, as temporal relationships are the most basic and most indispensable relationships.

5. Some philosophers, and theologians especially, try to define or explain time in terms of other things, as if time is not the fundamental feature of reality, as if the universe can be reduced to or is better explained by something more fundamental. My premise is that time is an absolute, the medium of all relationships and, indeed, the medium of our being and dynamism of our existence. Though one must remember that our experience of time is the multilayered and sometimes mysteriously interactive and interdependent product of the mind's struggle to make sense of and impose structure on time. The mind is ever attempting to map and navigate time. When the mind stops we experience an illusory sense of timelessness.

Timelessness is an illusion fostered by the perceived endurance or duration of specific things over long periods of time coupled with and spurred by our hopes of escaping destructive change. We want the things we value most to endure and in some sense be permanent features of our reality, as Plato, full of hope, believed his forms were. Those who nurture these hopes fail to appreciate that time creates value by creating uniqueness which requires emergence, duration, change, conflict, and destruction. Just as nothing outside of time could have relationships to anything, nothing outside of time could have value or meaning. Time is the great value creator—and destroyer. Value creation and destruction are the selfsame temporal process. Things are valuable only because they can be, will be, and are transformed and destroyed. By their transformation and/or destruction they create novelty and scarcity, two prevalent measures of value.

6.Time is the dynamism as well as the medium of reality. Time assumes the functions of Aristotle's unmoved mover, except time is pure movement that, according to Whitehead, advances in a creative *SPIRAL*, as distinct from a line or a circle. Indeed, the spiral combines a line and a circle, being both yet something different and more dynamic. Time, in being the medium of reality, is the medium of the immanent or pantheistic Actual God who creates reality in a spiraling process, the *creative continua.*

The concept of time as a medium of reality and a spiraling dynamic of its creation is difficult to grasp. Many prefer simpler views of time. Science tends to see the movement of time as linear like an arrow, while the ancients believed time moves repetitively in a circle like the seasons. Nietzsche's idea of eternal return reflects the ancient view, as does most classical Greek and Eastern thought. Circular motion repeats itself in an uncreative way, linear motion goes somewhere but leaves everything behind forever. The spiral image, as Whitehead saw, better conveys time's dynamic that repeats itself with significant variations wherein The God in Time continuously creates novelty and with it new value and meaning.

God does not exist outside of time: God is the creative spiral of time whose icon is written across the universe in spiraling galaxies, including our home the Milky Way. God is not within or a part of time because *Time itself is the dynamic manifestation of God. Time as God and God as Time is the absolute reality, the ground of meaning and the ultimate value creator.* Here I assert my own views, not those of Jung or Whitehead who were reluctant to venture this far. Eternity is not separate from time, eternity is the spiraling entirety of time. As such it cannot be measured or directly experienced in totality by consciousness within time. Eternity is the entirety of time, yet paradoxically it also resides in the present instant. All consciousness depends upon and subsists within time in the present instant. A god outside of time would lack a sense of the present and thus could not be conscious. Such a god would have no relevance to or impact upon anything within time, which includes humanity.

7. Time is like a field, a region in which a particular condition prevails, that condition being continuous interrelated, interdependent change. Time itself is the hyperfield or medium in which all change and indeed all that was, is, or will be exists. Just as there are variations in the fields of space, caused by gravity and other forces, there are variations in time fields caused by active archetypes, a crucial factor we will take up later. It has been argued that time itself does not exist or is unreal. Certainly, it is not a thing like other things. But rather than being unreal,

time, along with space, partakes of and defines all reality, since things are defined in relation to and by their positions within space and time, that is within their temporal fields or space-time. Like reality (and space), time is essentially relational; time is the great relator of all things past, passing, and to come. Try to imagine a thing that exists but is outside of time and has no temporal relationships. One can speculate conceptually, but when you try to get down to particulars, you are taken back into time. The ultimate reality, the supreme existential fact and challenge of life, the one thing we cannot avoid, or evade and must ever contend with is Time. And that makes Time Actual God. Actual God as the spiraling, creative dynamism of time and as the medium in which God acts are two sides of the same coin.

8. Time entails the movement and energy of an ever changing continuum of events. But the events, or the moments we perceive, are effects of inborn and cultural-linguistic structures the mind imposes on perception, the reality is simply the continuum. Movement creates space and mind fills it with light. The light of consciousness illumines time. There is more to consciousness than just its own light, consciousness partakes of all the senses, sight, sound, smell, feeling, and the intangible senses sometimes called intuition or insight. The light of consciousness in other persons also illumines time for each of us; however, it is more like a reflected light than a direct light. Consciousness can only exist in time; it arises with the discernment of relationships in time. All that consciousness can be aware of exists within the fields of time. Time, we assume, must exist prior to and independent of any specific, individualized

consciousness. The alternative is solipsism, the theory that only the ego exists, which we usually dismiss as insanity.

9. Time is the supreme creator and destroyer. The flower that once has faded forever dies. We can never return to the same moment in time, though often we can return to almost the same place in space with minimal apparent change. Through memory we can return to an image of a place or a period in time. Memories are traces in the mind that persist only as long as the minds who remember them. There is no assurance that the mind will not change or shift a memory over time. It usually does shift, so we keep diaries, photos, records, and mementos.

Time's fatal irreversibility creates longing for lost joys and opportunities, and dread and shame of past defeats along with fear that we are fated to replicate them. Romantic longing is the stuff of song and story, a bitter sweet aftertaste of happiness irreparably lost in time. Memory is our way of resisting time the destroyer, or retaining an aftertaste of our momentary experience of meaning, value, joy, pain, and beauty. In memory the past not only survives in the present, as Bergson noted, the remembered past provides a framework that gives meaning and value to the present.

Objects from the past can help us recall memories of feelings and sensations long departed. Proust's taste of madeleine retrieved a thread of memory to activate a flood of associated feelings and sensations. But it could not restore the original feelings and sensations themselves, only impressions of them. Our memories are like Cezanne's paintings of *Mont*

Sainte Victoire. Viewing the paintings is quite different from looking at the actual geological formation. We are viewing or sharing in Cezanne's remembered impression of the formation. A memory is more like an impressionistic painting than like a photograph. Memories are enlivened by consciousness through art. A memory is an artistic creation. Indeed, memory makes artists of us all.

The art of memory is essential to conscious thought, it engages the imagination, precedes and is more basic than logic or reasoning. The cave paintings in Spain, France, and the island of Sulawesi are among mankind's first steps into a new world of culture and structured, non-instinctual thinking and feeling. In the development of consciousness first comes art as remembered perception, then comes meditative reflection. All reflection is bound in time; all reflection involves, directly or indirectly, our endeavors to contend with time.

10. Clocks and watches measure time mechanically by the actions of pendulums, springs, or batteries moving dials or digits in space. In Jonathan Swift's *Gulliver's Travels* the Lilliputians speculate that Gulliver's watch must be a kind of god because he was always consulting it. Indeed, for modern man watches and clocks are like talismans or rosaries that connect us with our veritable modern deity, linear mechanical time.

But inner feeling or intuition, not the thinking coupled with sensation that connects us to watches and clocks, is our primitive and primary measure and our most authentic connection with time past.

Intuition is our sense of, our way to apprehend, the future which is usually colored by feelings of hope or fear, expectation, or consternation. Clocks and watches lock us into the present; we need inner feeling to apprehend the past and to reach out to the future. We need intuition to see time whole and put events in context. The importance of intuition in respect to time is a conviction that Jung, Whitehead, and Bergson shared.

11. Henri Bergson observed that we understand duration intuitively, not intellectually. Duration is not something we can perceive by the senses, since we directly perceive only the instant which is without duration. The sense of duration comes with the intuitive realization of the continuum or process of passing of time. Indeed, our awareness and knowledge of time itself is intuitive. Intuition allows us to access all three tenses and correlate real time and dream time to grasp the whole process of time. The intuition of duration forms the basis for memory. The fundamental things of human life do not change as time goes by, or so we hope. We long for and hope for duration. Without duration time would be a flux of unrelated instants. By enabling us to apprehend duration, intuition allows us to grasp meaning, value, and relationships, to discern what is fundamental or fleeting in human life. However, in today's world of rapidly escalating change fewer and fewer things seem fundamental to human life. It is as if duration itself is under siege. Are we headed for a vast evolutionary breakthrough where humanity creates its successors by adapting the core features of human nature to the challenges of strange new worlds beyond Earth? —a question that artificial intelligence, AI, may eventually make urgent.

Henri Bergson in his study

12. The world will always welcome lovers as time goes by, so goes the song. Emotions such as love and the drives to achieve and maintain identity and expand awareness of the world, life, and the self, are characteristic of developed consciousness. Fear of defeat, deterioration, and death, as well as hatred of injustice, seem equally fundamental to those blessed and cursed with the consciousness that brings a rich, complex awareness of duration in time. As consciousness expands so do our intuitions about duration and the processes that endure as time goes by. The more fully individuated our consciousness, the more we understand what endures for us, how, and why.

13. Technically, a conscious, decision making ego exists only in the present because consciousness itself only exists in the present. However, an individual self, which is the true subject as well as the ground of the ego and other psychic functions, is not limited to a single point of time, but it subsists in the unconscious from the hazy dawn of awareness in the womb until death, and perhaps beyond. One might say that self operates in dreamtime while ego operates in real time. To give consciousness depth and context we must exist in the present instant and subsist or endure across a span of time, we must have a perceiving ego that

can decide to act in the present but is grounded in and regulated by a durational self.

You cannot point to one moment in a life which represents the essential, enduring person any more than you can identify one cell in the thinking brain that controls the entire person whose wholeness consists of the relationships between his/her parts but is greater than any of the parts. Thinking, including self-consciousness, is a process, a *creative continuum*, wherein thoughts are distributed in an extraordinarily complex fashion throughout and generated by billions of nerve cells in the brain. A thought is never locked into one moment or one place. We think with our whole mind, though not our whole brain since much of brain function only regulates the body. Conscious thought may have originated as a guiding mechanism for creatures whose minds and lives became too complex to make decisions directly by instinct. Human consciousness, through the creation of language and culture, matured into something truly revolutionary with the power to change the course of life on earth and to create collectively veritable gods.

14. The generic word for the process of an individual subsisting across time and throughout the body, without being limited to any single time in life or place in the body is *consciousness: a unifying and coordinating force behind the entire interdependent and interrelating processes that form and maintain across time the sense of identity, as distinct from the perceived world, of the individual person.* Consciousness makes possible the discernment of temporal relationships including the intuition of duration, as distinct from the flux of change. Animals are alert

to change, but their discernment of temporal relationships and perception of duration are less refined than man's. Their minds are more directly attuned to their immediate experience.

Intuiting duration, human consciousness can transcend time as a succession of events without going outside of time or into dreamtime or a theoretical eternity. The very awareness of time as duration requires an intuitive transcendence of time that encompasses present, past, and future in a single vision. Men seek in transcendence of time a perception of eternity, but they are not the same. Transcendence of time is a condition for conscious distinction between past, present, and future. It is essential to intuiting duration across time and for unifying and interrelating the three interconnected tenses of time. Eternity, as commonly conceived by most religions, is a speculative escape from past, present, and future.

MEDITATION II: TIME AND CHANGE

1. Time and change are interdependent aspects of the same process: change references an action, time is the relationships or structure of a course of actions. Like change, time is a concept fraught with specific questions and general confusions. You can never step in the same river twice, said Heraclitus citing one of time and change's paradoxes. Like time, the very nature of a river is to flow and therefore change, otherwise it's a lake or a swamp. The river remains essentially the same process so long as it flows in about the same volume of water along about the same bed at about the same temperature. If it did not change or flow, if it ceased to be a process, it would become an immobile glacier whose temperature is absolute zero, an H2O rock not a river. Similarly, time would not be time were it to stop, all motion frozen in a single changeless moment, as some imagine eternity to be. Time never freezes, it changes constantly in its appearances to the observer. Yet the essential nature of time as a dynamic process of directional change does remain unchanged. Time is the unchanging presence of the interrelated processes or continuum of change directed across its three dimensions of past, present, and future.

Changeless time is impossible, or at least unverifiable. Everything that we can experience, lies within the process of time. (Likewise, modern physics has determined that there are things going on in "empty space" which means space itself has qualities that make totally empty space impossible. Time never stops changing and space is never empty. But space is not the same as its contents, and time is not the same as its changes. Time and space are mediums.) Change requires time to change in. To be outside of time and change is simply to not exist in any real sense. Even if there were nothing in a specific space or time, gravity, light, electromagnetism, and forces peculiar to space would still be interacting with that space or time; the general context of that particular space or time would be changing. Moreover, objects that appear unchanging, alter internally at their molecular and quantum levels. Thus, no object is truly unchanging. The foreverness of diamonds, like their beauty, is in the eye of the beholder.

2. A universe without time and change would be without relationships, differentiation, light, life, or consciousness, and therefore without value or a God. Light moves thru time-space creating a measure for change, the processes of life require constant change. The end of change is death. Consciousness exists in the present moment but consists of either reflection on how or why things changed in and from the past, or anticipation of how or why things will/may change in the future. We cannot conceive of a universe without time because consciousness is fluid, and it cannot exist without differentiating between things within the web of ever changing relationships that make up time. Time and change are the constants essential to any universe that could contain mind and consciousness.

3. Change allows the measuring of time. while relationships, including those of duration, show the structure of time. Time sustains existential reality. Yet conscious individuals long for eternity, a state beyond time and without change, not realizing such a state would not allow for consciousness, truth, value, or meaning. Those who long for eternity believe it would be bliss, unending happiness, nirvana. However, it would simply be nothingness, because consciousness cannot exist in a state without time and changing relationships. Eastern meditation techniques minimize our awareness of time and change, but they cannot eliminate it as long as we breathe and our hearts beat. They lower our threshold of perception of change, true solitude does the same with less effort.

4. Eternity means either all of time or something unknowable beyond time and change. In the latter sense it can be viewed as the nothingness against which men attempt to define the ever changing realness of time. But eternity does not exist except conceptually, it cannot be even imagined except as an abstract absence of time, change, life, consciousness, and existence. The nothingness of eternity beyond time cannot be good in itself because the good requires time and change to allow for conscious beings to perceive, create, and enjoy the good. Consciousness allows us to see how time with its ever changing relationships gives events value and meaning.

5. Like change itself, the complexity of attempts to transcend change is confusing. So let me try to summarize and simplify. Men and women, striving for the meaning they fear is absent from their lives, hope for an eternity that does not change or has certain unchanging qualities,

like eternal bliss. They fail to see that there can be nothing real beyond time because nothing real is changeless; except that nothingness itself is an abstraction, like the number zero, that does not change. Neither the physical nor the psychical can exist without time and change. If a thing seems not to change that is an illusion: its temporal context of relationships is always changing and that temporal context defines what anything is. All things exist in ever changing, infinitely complex interrelated temporal contexts within the vast, fluid web of time. All men and all things are born in a temporal context and there they grow, change, endure and die. Time as change is the absolute context of everything and everyone, in this respect also Time is Actual God

.

MEDITATION III: TIME, GOD, AND MAN'S RELIGIONS

1.Ideas about time and God develop interdependently with consciousness itself. The realization of identity that makes us aware of being different from the world and from each other, brings with it a recognition of the ways the world is subject to time. In an attempt to make time's creative and destructive powers more auspicious, if not more comprehensible, we search for gods. A few serious thinkers, such as Jung and Whitehead, ultimately search for a God who acts in time the way the self acts in the psyche, as a homeostatic regulator. In the search for gods we initially construct god images that, according to Jung, derive from the god image or archetype in every psyche but become culturally specific and, therefore, distinct from the archetypal god image, just as the ego is distinct from the self. However, men conflate, or to use Jung's term 'contaminate,' ego with self leading them to conflate their god images with a metaphysical entity, Actual God.

Human egos and our god images arise in part in response to the insight that we are separate from the world we perceive but do not control. However, the development and character of both human egos and our god-images is shaped by the cultural context that provides the values, language, and conceptual framework for talking and thinking about them. Like the self, Jung concluded, Actual God is a hypostasis, or an underlying reality, not a cultural construct. Philosophers ponder the nature of the self and God as they as they explore the world's interconnectedness, puzzle about passing time, and confront the disturbing, yet thought provoking, realization of the time's irreversibility and the inevitability of death. From such speculations is metaphysics itself born.

Jung's scientific training inhibited him from venturing into metaphysics. He wanted his psychology to remain empirical. To avoid the questions of metaphysics Jung, unlike Whitehead, avoided speculation about Actual God. Perhaps that was why he never remarked on Time as a creative and ordering force being Actual God manifest, even though his interest in synchronicity might have led him there. His reticence may have limited the influence of his psychology in two related areas whose clinical import deeply concerned him, theology and philosophy of religion. In contrast, this work will at times venture into places where Jung, and Whitehead, declined to tread.

2. What shapes individuals, their priorities and their cultural context, will by analogy shape their god or gods. The term anthropomorphism refers to the practice of making gods in a human image. Looking in the mirror, hubristic humans imagine they see an image like the God of the universe when they are only contemplating the image of a biped primate that has existed for less than one percent of one percent of the span of time scientists attribute to the universe. Atheists argue that

men create gods in their image, while anthropomorphic theists say that God created man in His divine image. Setting their contention aside, we can see that Actual God is a process far greater than humanity, a process that unfolds, in the universe across time as the self does within the individual. To put it another way, Actual God is like the homeostatic self and transcendental oversoul of the universe.

Jung's God at times exhibits the metaphysical character of an oversoul, and Whitehead's God certainly does. But Jung the clinician focused on the personal God-image, that is whatever his individual patient, experienced as God. He saw that a force emerges when god-images become energized, Jung's term for this force, which unites opposites to further consciousness, is the *transcendent function.* It activates compensatory elements of the unconscious that allow us to transcend destructive, one-sided attachments. The transcendent function lets us to move forward by creating a dialectic wherein opposites dialogue with each other and thereby overcome the dualism of opposites locked in perpetual conflict. The Paraclete, or Holy Spirit in the orthodox Trinity, manifests the work of the transcendent function within the Christian Godhead.

Carl Jung in his study.

For Jung, God, like the psyche and reality itself, is uniplural. Indeed, he saw the Judeo-Christian Godhead as a Quaternity reflecting the four-part wholeness of the psychic and the three-part dialectic of transformation and individuation. Jung's view can allow for sub-gods, beside the transcendent or actual God, or for many viceroys to the oversoul of the universe (a Mormon view, some speculate). From this perspective, the highest god humans know and experience most frequently, directly, and fully might simply be the god of nature or life on Earth, which the Greeks called Gaia. Such a god (or goddess) would be greater than a tribal god or even a god of a civilization like the Christian Godhead or Allah. It still is not the Actual God of the entire Universe with all its spiraling galaxies and myriad mysteries to match. Actual God more resembles what Aristotle called the unmoved mover, what Hegel called the Absolute, what Alan Watts termed the Supreme Identity, and what this work identifies as The God in Time.

Skeptics might dismiss as fantasy a civilizational god, like Allah or the Christian Godhead, formed by collective psychic energy rising from distinct human communities. A collective god for a human group, culture or civilization, however, is fully credible, indeed commonplace. The collective unconscious, as distinguished from the personal unconscious, is a repository of archetypes common to all mankind. As such, it is the source and equivalent of a vast polytheistic pantheon whose members are projections of aspects of archetypes. The ego, shadow, anima, animus, persona, and the mother and father, the brothers or twins, and other archetypes join to create a congress, or society, of the individual person. Each archetypal function has its own unique character and distinct role in coloring and shaping the identity of the person we think of as a single individual. Just as each person forms his/her personal identity from a collective of voices or archetypal functions coalescing within the self, advanced cultures, or civilizations, form a divine identity or a Godhead from the collective of dominant archetypes coalescing within their culture.

In this view, the cultural deities are personifications of the society of archetypes that are energized in specific cultures. The realm of computers affords a rather loose analogy to this identity formation among cultures and human individuals. Linking large numbers of smaller computers in parallel can achieve computational power more economically than giant single machines. One might compare the small computers connected in parallel to the active archetypes forming a deity within a culture—or just to human teamwork. Teams are powerful instruments that allow their members to get results they could never get by themselves. We see this in nature, in packs of dogs or lion prides for example. Human and animal societies are teams; humans, dogs, orcas, ants, bees, and termites all succeed by teamwork. Similarly, the bodies of complex animals consist of groups of specialized cells working as teams on specific tasks for the benefit of the entire animal. The human psyche is hardwired to understand the value of teamwork so that we have a propensity to cooperate with the drives of societies and cultures to form their own identities independent of the individual members. However, the ego's obsession with dominance, the ego is the alpha dog of the psyche, can blind us to the teamwork among archetypes that is essential to psychic stability and wholeness. Split personalities, like Dr. Jekyll and Mr. Hyde, display the pluralistic psyche that our alpha dog egos are reluctant to recognize. Self-awareness comes when we realize the ego is not all of me, it's just the quarterback of my psychic team. Similarly, a cultural Godhead is not always dominated by single archetypal function like the ego, but is more often a team of rival archetypes: consider, for example, the Hellenic and Hindu pantheons.

3. I will use the term *Archetype* to refer to a range of psychic images, patterns and nexuses of energy arising from common human experiences and reflecting inherent neuro-psychic structures. Archetypes may involve totems, symbols, themes, ritual acts, basic situations, or figures that have meanings understood by an entire culture, or all of humanity, such as the mother and father. Archetypes recur in myth, literature, folklore, the dreams of patients, theology, and even in politics. They can be identified by their peculiar resonances and power to raise affect.

Archetypes act in tribal groups, nations, and entire civilizations as shared or fused psychic energy, they do not act just in individual psyches. The activity of archetypes is an often overlooked force that can help explain crowd behavior and group psychology that otherwise seems inexplicable or irrational. For example, Jung surmised that a revival of the repressed archetype linked to the German god Wotan explained much of the dynamism of the Nazi movement, and of Hitler who became, in effect, Wotan's avatar. A positive instance of archetypal activation is the Christ archetype giving dynamism to the work of devotees, such as St. Francis or Mother Theresa, and energizing their followers. For better or for worse, people under the sway of an archetype think and react as a group operating under crowd psychology.

Jung identified two principle forms of archetypes:

1) personified archetypes such as the anima, animus, persona, shadow, the mother, the father, the brothers, the trickster, and the wiseman or wisewoman;

2) archetypes of transformation which involve a basic situation or pattern like birth, *conjunctio*, initiation, sacrifice, marriage, and quest. Although archetypes appear embedded in images, they themselves are psychic predispositions to form, modulate, and organize images. Archetypes differ from the instincts of animals in that men, having the gift of superior reflection, can more readily oppose, tame, direct, shape, and transform archetypal energies. Furthermore, archetypes combine and interface so that manifestations of particular archetypes are always modified by other archetypes constellated in the individual's psyche and

the cultural milieu. For example, the shadow will become Satanic in one manifestation, Promethean in a second, Mercurial in a third, and a trickster or merely clownish in a fourth. Archetypes usually have potential for good, bad, and all the shades in between. Jung believed that the psyche constellates in dreams and in myths, myths being the collective dreams of a culture, those archetypes that compensate imbalances in the common beliefs of the culture. The homeostatic function of the self-activates archetypes, most often through images, to promote adaptation, connection, growth, and healing.

4. We can communicate with our fellows, but we cannot communicate with the individual cells of their bodies or even of our own bodies; whole communicates to whole, part to part. The same analogy may hold for our cultural gods and for individual humans, since gods and individuals are members of different orders of organization. The gods are products of the interrelationships within groups of people. Similarly, individuals are the product of interrelationships among groups of cells, and their personalities are products of interrelationships among psychic functions or archetypes. The inability of individuals to communicate at will with specific cultural gods does not, then, prove that these gods do not exist. A cultural god normally communicates with humans as a group, most likely though channels of crowd psychology, spiritual intuition, and in myth, though there appear to be instances of direct communication with select individuals.

Individuals can communicate directly with each other when they subsist in overlapping timeframes. Indeed, all individuals subsist within single, limited timeframes, which we call their lifetimes. A culture's pantheon subsists in a timeframe that spans the life of the culture. An exception is the Goddess of Life on earth: Mother Nature subsists across geological eons. Actual God's timeframe is even larger, it must encompass all moments throughout time, yet transcend, every single moment.

"Know" can mean and involve many things that do not require direct perception; for example, intuition is indirect knowing. The possibility that we can know Actual God intuitively is real. However, it may be no more reasonable to expect to communicate directly with Actual God than to have a conversation with gravity. Nonetheless, one must learn to live in harmony with gravity. Falling at odds with it can be a crushing experience. Likewise, we must learn to live in harmony with Actual God, Time as God.

Jung appears to have believed that we can know the existence of Actual God intuitively. I say "appears" because the distinction between the personal god image, the cultural godhead, and Actual God are not always clear in Jung. Each of Jung's four psychic functions, thinking, feeling, sensation, and intuition, is a type of "knowing," and humans have claimed to know God, or at least some God, in each of these ways. For Jung, God is real in the sense that some ideas, like the idea of God, are intuitively common to humanity. Actual God, the One Who creates and sustains the universe, is such an idea. Nonetheless, Actual God, or even specific cultural gods, remain only partially knowable by individual men because their consciousness is of a different order. We are a part that cannot, by definition, communicate directly with or fully know the whole, especially the whole universe and its God. Nonetheless, Actual God can and apparently does sometimes communicate in indirect or veiled ways with us, for instance in mystical experiences. Most commonly this is experienced as a sense of the meaning, value, and beauty in life and the universe.

5. The God or Goddess of life on Earth, may communicate with individual species and even individual creatures just as the wholeness of an animal may communicate with its cells. Indeed, just as the regulatory self, as conceived by Jung, communicates with the parts of the psyche, especially the conscious ego. One definition of self is the essential wholeness, or unifying dynamism, of an organism. Complex organisms would, then, have more complex selves. When we speak of the unifying dynamism of a society or culture, we use the related term soul, rather than self. The God or Goddess of Life on Earth may be seen as the soul and the self of the biosphere.

A culture or nation can have a soul metaphorically speaking. We talk about the soul of a nation as that which unifies it. Group souls function as god images for the group. Group souls may under certain conditions communicate with the group or with individuals in the group. When people or their leaders claim to be communicating with or following a god, they are often talking about their group soul. Apotheosis of a group soul is a way tribal and cultural gods are formed. Such deities can exist as long as their cultures which may be for millennia. They can act like independent spirits, even as ideas take on a life of their own, for better or for worse.

6. Nietzsche is famous, or infamous, for declaring the death of God. Cultural gods can fade and even die, though their light may span millennia. They are born out of and sustained by the archetypal energies behind people's beliefs. When the fires of belief no longer blaze forth, their light fades and at last extinguishes. The fires lighting the gods of the ancient world, Zeus, Hera, Apollo, Dionysius, Aphrodite, Ra, Isis, etc., have all died, and men now see in them only archaic relics. Deceased gods tend to be dismissed by the living as tribal or cultural instances of collective *folie a deux.* These are fantasies shared and sustained by groups who accept them as true representations of reality because of reinforcement from their mutual beliefs. Instances of crowd psychology applied culture wide across great expanses of time are common and not confined to religion. Nonetheless, crowd psychology flares tend to be short lived, and crowd psychology is not in itself an adequate explanation for long term belief in a god. Powerful archetypes stand behind the long lived gods of the ancient world making them potent psychic, cultural, and political realities. As was often the case with Nietzsche, he over-simplified to make a point.

Nietzsche's aphorism referred particularly to the decline of belief in the god images and theological concepts associated with traditional Christian morality, which he criticized for inhibiting human creativity and impeding full development of the self. This decline of traditional Christian morality is relative. Although certain denominations of Christianity are in decline today, others stressing traditional morality are thriving, and some competing religions show signs of vitality. Visiting India, I was impressed by the vitality of the Hindu religion; the same for Buddhism in parts of

Asia, and Allah burns bright wherever his true believers abound. Moreover, totalitarian political movements that deify their own leaders, ideology, and morality, thereby replacing the older religions, have flourished in the post Nietzschean era. Though much to the detriment of mankind, many would argue.

7 A strongly held group belief powered by constellated archetypes functions like a god even where it is not attributed to a specific deity. Some say that greed is the god of capitalism and consumerism its morality. Jung thought Freud had made sexuality his god. Similarly, power lust is the god of totalitarian systems like Nazism and Communism. The Greeks personified every basic human drive by attaching a god to it. Thus, war and strife became Mars, desire Aphrodite, reason Apollo, wisdom Athena etc.

However, Greek philosophy eventually identified deity as the ordering force of the universe, a step which favored monotheism over polytheism. Christian theologians knew that their God had to be associated with some appealing aspects of humanity to preserve the active interest of common humanity. Moreover, manifestations of the divinity needed engaging human images. Jesus, the humble Jewish carpenter who was believed to be the Son of God willing to sacrifice himself for man's sins, and his lovely virgin mother Mary, effectively served this purpose for nearly two millennia.

The genius of Christianity was to make Greek philosophy's abstract God of the universe, the prime mover and ground of truth and reason, directly intervene in human history through Jesus, an actual person whose character was noble, gracious, and heroic and whose ethical teachings embodied enduring wisdom. But over time the other accumulating divine interventions within Judeo-Christian scripture, myth, and lore began to ignite a skepticism that culminated in David Hume's *Dialogues Concerning Natural Religion.* With the spread of scientific thought like Hume's, the miracles became millstones. Some notable interventions involving violations of natural law that were especially difficult to swallow were the creation of the world/universe in seven days six thousand years ago, Adam, Eve, and Satan in Eden, Noah's flood, the sun stopping for Joshua, the virgin birth, miracles attributed to saints, and Christ's physical resurrection.

However, the credibility of Actual God, a force that transcends history and humanity and orders the universe, e.g. Time as God, remains difficult to challenge and impossible to refute. Nonetheless, linking Actual God to the hopes and biases of a tribe, or a civilization, or even a single species is a major stretch. It requires a leap of faith. Post Nietzschean moderns had to choose between an impersonal, remote and seemingly theoretical deity who presided over the cosmos, and a gaggle of cultural gods who require suspension of disbelief in respect to the truth of their adherents' claims. Post Nietzschean intellectuals are reluctant to make religious leaps of faith. Regrettably, the same reluctance does not always apply to ideological leaps of faith.

8. So what may we conclude of Actual God as a culturally non-specific concept, the one God beyond the plethora of anthropomorphic god

images, the God of Whitehead, Shakespeare, and the mystics, and possibly Jung when in a metaphysical frame of mind, the God of the cosmos who cannot be blindsided with positivistic skepticism or Nietzschean moral sarcasm?

First, God, either as a universal deity, or as Nature the subgod or goddess of all life on earth, requires time and process to unfold. Actual God unfolds meaning and value in the world through time and process; the self and the soul do the same within the individual person. God is the life, the structure, and the intelligence of the temporal process, the *creative continua*, that forms the universe we know. The two words that best describe what we know of God and why are time and life—both are processes. God could not exist in a static universe because there would be no temporal process in which life, value, truth, beauty, and meaning could unfold.

Second, the universal deity cannot be perfect and eternal as in the claims of orthodox theology. A perfect, eternal, unchanging deity would make the universe static and time an illusion which is counter to all our experience. Meaning, value, and life require a beginning, development, and an end with a potential for rebirth in a child, resurrection, or perhaps reincarnation. For common humanity meaning follows the life cycle of birth, growth, maturity, decline, death, and rebirth.

Third, the universal deity cannot be omnipotent and omniscient as that would deprive his creatures of free will, adventure, creativity, and responsibility. All good things for humanity would be sacrificed upon the alter of the deity's power drive. Where power rules absolutely it corrupts, but first it drives out love. As Whitehead, observed, God's love requires that He limit His power by giving man a measure of independent creativity.

Fourth, the universal deity must afford an intellectually, and spiritually satisfying account of the salient characteristics of the human condition and the universe as we know it. We cannot ascertain the ultimate laws of the universe or its God, since the universe and its God are unique, while laws, in that word's scientific meaning, must cover a range of similar phenomena and be disprovable. However, we can talk about salient characteristics. Uniqueness or novelty is a pervasive characteristic of the universe. Not only is the universe itself unique, but everything in it is unique in that above the quantum level everything and every event has its own singular place in space and in irreversible time. Even things that otherwise appear identical, like identical twins, have their unique position in the web of space-time and, therefore, different relationships to everything else. Uniqueness is an essential aspect of value, and the way that The God in Time, creates value. The God in Time is the value creator; as Whitehead indicated, creativity is this God's defining activity and character.

9. There are five principle stages or meanings for the term "deity," or five faces of God, ascending in a hierarchy from the personal God images to the pantheistic Actual God of the entire cosmos. These stages reflect degrees of universality. The lower stages are usually what people refer to as "God." They are not the same as the Universal deity or Actual

God, even though people naively assume they are referring to Actual God when they are really talking about a lesser entity on a different, more limited plane.

*I.*THE GOD IMAGE IN THE PSYCHE. A fundamental tenet of Jung's psychology of religion is the presence, and experience, of a god image or archetype in the psyche of each person. Jung wrote: "I know the existence of the God image in general and in particular. I know it is a matter of universal experience." All theories of deity are based on and arise from this universal experience. Why, Jung and other theists will ask the atheist, do we have a God image unless it refers to something and has a function? Theirs may be a psychological version of the ontological argument for the existence of God and may fail to persuade atheists. Yet being psychological, it has a basis in verifiable experience and cannot be dismissed as just metaphysics. Nonetheless, most proponents of deity tend to move quickly from the realm of archetypal experiences into theological speculation as they attempt to define deity itself in terms familiar to their own cultures.

*II.*TRIBAL AND CANONICAL INTERPRETATIONS of the deity are versions of or perspectives upon civilizational Godheads: in particular, the official tenets of Catholicism, Judaism, Protestant denominations, Greek and Russian Orthodoxy, and Mormonism for the West. Islam, Hinduism, and Buddhism also present diverse sectarian conceptions of their civilizational Godheads. For example, Chinese, Thai, Japanese, and Tibetan forms of Buddhism differ significantly. In all the religions devotees begin with this lower conception of deity and most are never encouraged to elevate their spiritual sights beyond it. Its limitations, therefore, can make them subject to loss of faith and disbelief leading to despair and recreancy.

III. CIVILIZATIONAL GODHEADS are to a civilization as the self is to an individual. They are human culture's most sophisticated instruments for dealing with Actual God. Godheads are formed as a civilization projects a society, Whitehead's term, or a family of archetypes upon Actual God. An example is Yahweh who formed from coalescence of several ancient god images shared by the Hebrew people. When the archetypes are activated in the population or individuals, they give power to the Godheads which become psychic entities that can act on their own and have their own distinct character and capabilities. Except for the philosophically minded, typical worshippers fail to make the distinction between the archetypes forming their Godheads and Actual God. Careless identification of a cultural artifact with Actual God has made easier the debunking mission of atheists.

Until Jung, Western thinkers paid scant attention to the archetypal forces behind our civilizational gods. In *Answer to Job* Jung maintained that man changes god, or that god, meaning Godhead, learns by interaction with man. This is true for the civilizational Godheads, though not of Actual God. Jung believed that the western Godhead evolved, from the Old Testament deity who tested Job, into a more conscious and ethical deity who wanted to fully understand what it is to be man. He lived the life of Christ and died on a cross not to pay the penalty for mankind's sins but to understand his creature's lot. (I would add that Christ also came to give man universal ethical guidelines and illustrate them through his life.) Jung stated that "the real history of the world seems to be the progressive incarnation of the deity." A more accurate statement might be the history of the Western world is the progressive revelation of the Judeo-Christian Godhead.

We neglect to study the archetypes behind civilizational Godheads. As a consequence, we fail to understand their influences on our civilization. In the Western Christian Godhead, the Supernal Archetypes are four: The Father or primordial parent; the bifurcate Two Hostile Brothers archetype represented by Christ-Son of Man on one hand, and Prometheus-Lucifer on the other; and the Anima (Animus) or wise person represented by Sophia or the Paraclete. *Answer to Job* shows how Orthodox theology turns the Godhead upside down placing the father at the top and neutering the feminine. Islam goes further in the deification of the male principle by eliminating all but the supreme Father, with Satan, the bad seed, exiled as a fall guy. The most accurate and vivid, though so far largely unrecognized, revelation of the true archetypes of the Western Godhead is found in a seemingly unlikely place: Shakespeare's supreme tragedy *King Lear*.

The full potential of the western Quaternal Godhead to improve society and lead men to experience Actual God remains untested. The Trinity, *summum bonum* and *privatio boni* doctrines impede the unfolding of the Quaternity and force it underground into literary works where symbolic revelations are protected by "artistic license." However, archetypes cannot be banned. The Quaternal Godhead is not dead, it will continue to manifest itself in unexpected ways throughout Western Civilization. Overall, the Judeo-Christian Godhead has been the most powerful archetypal force in human history. It promoted political and personal freedom and ended slavery. It's view of time as a process with a stable past and a dynamic future facilitated scientific prediction, spurred Western technology, and made progress a viable ideal. Today mortally challenged by totalitarian ideologies, it may yet triumph bringing new benefits to mankind.

Archetypes operating powerfully within a culture or a subculture of a civilization can be outliers from its Godhead. They can function as counterforces, qualifiers, and correctives to the main Godhead. Some examples in the Western context are the old Germanic and Norse gods Wotan, Odin, Thor, the trickster Loki, and Nietzsche's favorite Hellenic divinity, Dionysius. Modern totalitarian movements and ideologies similarly operate as outliers to the Western Godhead.

*IV.*GODDESS OF EARTHLY LIFE, sometimes called Gaia, or the oversoul, represents the archetypal life force active on planet Earth. The Greeks revered Gaia as the primordial Mother who nourishes and cares for all life. According to Hesiod's Theogony all the gods originated from her. Kali is Gaia's more formidable Hindu parallel. The Goddess, Mother Nature, harmonizes nature and keeps various life forms in an overall symbiotic balance. James Lovelock has re-envisioned Gaia in scientific terminology as a feedback system that sustains the biosphere. The TAO has a similar homeostatic function and goal. Taoism is a way of life whose objective is balance and harmony with nature. Taoism is the most prominent traditional belief system that centers on nature as deity. Its focus is on the individual's relation to nature. It contrasts with the Confucian stress on the individual's relation to society, and the Western and Islamic stresses on relation with an egolike deity requiring obedience to specific laws. Setting aside its compromising political features, the ecological movement in modern societies expresses the archetypes of nature that inspire and infuse Taoism and have had wide impact on Japanese culture.

V. ACTUAL GOD, or the god of the universe, is inherent in the fundamental characteristic of time: irreversibility and its corollaries creativity, novelty, and interdependence. Irreversibility makes possible truth, because the past cannot change; value, because events and things are novel; and meaning because conscious beings are unique and free to shape whom they become and are therefore responsible for who they are. The universe is indisputably ordered by time making Actual God irrefutable, a view that accords with Whitehead, and is dramatized by Shakespeare. Jung in the Red Book gives Abraxas a position like that of Actual God, and later refers to Abraxas, the Gnostics supreme deity, as a "time God." Later Jung, the survivor of two world wars, asserts only the God image, withholding judgment on its metaphysical corollary, Actual God. Zen Buddhism can be seen as a path to direct experience of Actual God, although Zen generally repudiates "belief" in God, and rejects belief in a personal egolike deity. "There are, indeed, things that cannot be put into worlds," said Wittgenstein in a Zen moment. "They make themselves manifest. They are what is mystical." Mysticism assumes God in known intuitively, and is like a self or a universal homeostatic function, a *unio mystica*, rather than like an ego.

Monotheism often arises from a direct experience, usually intuitive or visionary, of Actual God. Moses and Mohammad are prime examples. However, history shows that in practice the direct experience of Actual God is usually displaced by dogmas that formulate ego based civilizational Gods. After Christ, some mystical Jews turned from Godhead to pursue directly Actual God with the Kabbalah which expresses the Jews efforts to move beyond tribal and civilizational gods to engage with Actual God. Christians developed further their civilizational Godhead in archetypes of Quaternity that reflect the self. However, with the doctrine of the Trinity, Orthodoxy created a tribal deity that subverted the Quaternal civilizational Godhead.

The four preceding levels of deity allow for personal gods; but Actual God, as presented herein, entails a form of pantheism that focuses on the awesome creative and destructive forces of time. It makes tribal and civilizational gods ultimately subordinate to Actual God, without denying their reality and significance for mankind. A definition of Actual God: *the dynamism that is and moves through time and by that means creates and sustains the universe. Actual God entails the creative interconnectedness and interdependence of everything in time, and thereby maintains cosmic homeostasis. In coalescing the unus mundus, and through time's irreversibility, Actual God gives value and meaning to all things and thus acts as a metaphysical guarantor of the validity of religious experience. Moreover, the irreversibility of time and the interconnectedness and interdependence of everything in time give Actual God moral qualities and provide a universal ground for ethics.*

10. St. Augustine' deserves credit for profound insights into the human condition, time and God. Time's irreversibility, he maintained, makes everything unique and gives unique value, or novelty, and meaning to each moment. Augustine believed that God lives in eternity but acts in time to create value. (He resides in a Hellenic universe, but works on the other side of the tracks in a Hebraic universe!) His two most important actions are the creation itself and His own personal intervention in time through the life of Christ. Time itself moves irreversibly forward making everything unique; thereby, time becomes God's engine of universal creativity. Augustine's dual view contrasts with the Platonic and Aristotelean conceptions of time as a revolving manifestation of static eternity.

Saint Augustine in his study, by Phillippe de Champaigne, 1660

The irreversibility of time creates the problem of evil, or theodicy, which greatly troubled Augustine and posed a challenge for theologians and philosophers ever since. God for Augustine and Christian orthodoxy is unchanging, perfect, and eternal which means He must be outside time. So why did He create a temporal universe abounding in irreversible evil and waste? Augustine insisted that the evil is only apparent, a deprivation of good, *privatio boni.* Of course Augustine lived too early to have profited from reading *King Lear, Othello, Hamlet,* or *Macbeth,* let alone known about Hitler and Auschwitz or Stalin and his Gulag.

Eternity is a metaphysical speculation, not a reality that we can experience. By recognizing that Actual God is not only in time, but is time itself, we remove, if not resolve, the problem of a good God creating a universe containing evil. The God of time is Self-creating as It/He/She/They create the universe through time. Actual God unfolds in that temporal process of creation of which tragedy, waste, and evil form unavoidable aspects. Because God and time, including all moments of time, are unique and there is nothing to compare them to, they and their actions cannot be judged by any standards but their own. The part cannot judge the whole since it lacks an Archimedean point outside the whole. Yet we directly perceive the universe, and God, since we are in it, that is, in time. We can see and grasp, although Augustine did not fully, that by creating uniqueness time creates value and meaning all of the time.

The claims by Augustine, and many other proponents of mankind's tribal/canonical religions, that God is perfectly good demean human and animal suffering, and deny humanity a morally credible God.

They give the creator a perfection that flies in the face of all we know in life and of the world. They drive God from existence into essence: that is, they drive God out of the universe we can experience into a hypothetical eternity that lacks existential reality. The God of Time is replaced with a power dogma based not on the self, which reflects Actual God, but on the ego's will to power and perfection: absolute perfection being an inflated claim expressing ego's love for absolute power. The proponents of a perfect God want to judge the values created in time by Actual God and the self, by their own limited ego standards. Their assertions that God is perfectly good, omniscient, and omnipotent, and that their religion's dogmas are ultimate truths, amounts to intellectual bullying: "it is true because we say it is true, like it or lump it." The essential religious attitude, and the opposite of bullying, is not willed faith, as with some Christians, or abject submission as with Islam. The essential religious attitude is a spontaneous awe that both humbles and exalts, and in the end inspires creativity.

11. Attribution to deity of will to power is a common pitfall of monotheistic religions. A grievous limitation of Islam, the strictest monotheism among religions with a mass following, is its implicit deification of ego's will to power and static identity, coupled to its explicit demand for perfect obedience and absolute submission to inflexible codes. Deification of ego stunts growth of the soul and the self which must unfold through the unpredictable processes of time in the adventure of life. It undercuts the value creation that is life itself. Dogmatic Islam and canonical Christianity make God a projection of the ego's power fantasy of being outside and above time. Indeed, for some Islamic theologians

their deity embodies of the fantasy of conquering time that inspires the belief that Allah can change the past through His absolute omnipotence. This can be viewed as blasphemous in that it makes God like an irresponsible, power crazed despot.

12. The purest, in the psychological sense, of the major, contemporary religions, Jung theorized, may be Christianity and Buddhism because they come closest to presenting the self as a reflection of God. Both have a limited process view of time and history. But the limitations are critical. Buddhism moves toward flux or process in its Japanese versions, but the Chinese and Indian forms are infected with stasis assumptions endemic to those cultures and the Hindu language, and to the conception of nirvana. Moreover, Buddhism's feeling grasp of the existential reality of time is restricted by refusal to allow unblinking recognition of time's irreversibility and its fatal implication, tragedy. While Christianity posits a process system during the course of history, Christian eschatology puts a stop to time with the last judgment and inserts a major break in time with the incarnation of the Son of God. Also, Christian orthodoxy's tenet of the immortality of the soul, borrowed from Greek Platonism, is at odds with the process view of time implicit in the basic Christian myth and archetypes of Godhead.

Islam's goal of political domination through systematic societal regulation makes it more like an ideology than are the other major religions. "Render unto Caesar" was not Mohammad's dictum, which was more like "convert Caesar to Islam." The Prophet and his followers strove to merge the political and religious realms in an unyielding theocracy under Sharia law. This has led hostile critics to dismiss Islam as a political ideology masquerading as a religion. Such generalizations refer to the theoretical aspects of Islam, not to what it means to individual devotees. Personal meaning varies widely and all the great religions have sincere devotees whose dominant motivations are more spiritual than political, social, or economic. All the major religions offer paths to god, however, convoluted, crabbed, inconvenient, and fraught with pitfalls.

.

MEDITATION IV:
TIME, GOD AND MAN'S IDEOLOGIES

1. Ideology arises from political ambitions spurred by economic desperation and alienation, as with Marxism, an ideology that demands religious fervor and total commitment toward the ends of political power, economic control, and social regimentation. Islam arose in an era when heavy trappings of religion were needed to make any ideology palatable. The initial Abrahamic religions, Judaism, Christianity, may be viewed as having generated two great Abrahamic ideologies. The first, Islam, is a religious ideology, the second Marxism is secular. They are two of the four dominant ideological forces in the current political era. Marx is famous for saying "religion is the opiate of the people." After Marx and to a large extent because of him, it can be said with greater truth, "ideology has become the opiate of the people."

The third major ideology, Capitalism, which stresses the individual where Islam and Marxism stress the group and define individuals through their membership in groups, owes its origins in part to the basic Christian and Jewish emphasis on individual freedom and commitment to the ideal of progress. These were given a major boost by the Protestant Reformation. Feudalism and monarchy, which stress a hierarchical social order, were long the basic ideological stance of Catholicism and Eastern Orthodoxy. Even today Catholicism retains its medieval hierarchy with a monarch-like “infallible” Pope at the top, and Cardinals functioning like an aristocracy. Christianity tends to generate a plethora of ideological permutations as it endeavors to implement in society the fluid, multifaceted teachings and social concerns of the religion’s founder, mediated through the variant notions of the founder of the church, St. Paul and its great thinker-teacher St. Augustine. Protestantism is less church oriented and less bureaucratic, more fluid, and more open to adaptation than Catholicism or Eastern Orthodoxy, which are, nonetheless, decidedly more elastic than Islam.

Protestantism, a religion born of the rising capitalistic middle class, encourages freedom and progress by allowing the individual a direct relationship with God, outside the mediation of a priesthood. This elevation of the individual, (“all men are created equal” is a characteristic Protestant insight,) prepared the way for capitalism’s sometime partner, democracy. The capitalists are correct in asserting that equality and freedom must include economic opportunity and freedom. However, when the economic freedom of the few trumps ethics, as it too often does in capitalism, the elite curtails freedoms for the many that are essential for a healthy, stable democracy. Unrestrained “free market” capitalism too quickly becomes ruthless opportunism which undermines the social contract. Restraints are necessary to preserve the support of a solid majority and, even more important, the human dignity of individual citizens.

A glaring contemporary capitalist violation is the endemic failure to effectively enforce fraud laws that protect the populace from exploitation and theft by powerful institutions and unscrupulous individuals. Fraud and a host of serious infractions are relegated to the civil court system where enforcement of rights depends on the fraud victim's ability to pay lawyers. "Equal protection of the law" becomes a cynical joke in a judicial system where pay to play effectively abrogates "do unto others." The problem worsens with the increasing wealth gap between ordinary citizens and powerful institutions and individuals. Two opposing classes of citizens result: those who can buy justice or injustice, and those who cannot and must suffer the consequences. The division bodes ill for the future of equality, freedom, democracy, and the Judeo-Christian ethic itself.

The fourth major ideology of modern times, positivism, arises in part from a rebellion against religion, or at least agnosticism about theological claims. Instead of turning to religion for values, positivism attempts to discover values in scientific endeavor and method, and substitute them for religion. Such terms as scientism, materialism, mechanism, and postmodernism mark the territory of positivism's ideological bent which assumes that every justifiable assertion can be proved scientifically or mathematically. Its more zealous adherents tend to flatly deny metaphysics and religion, devalue art, leave little room for intuition or emotive understanding, and abrogate the soul. While positivism and its materialistic cohorts are scientifically oriented, many great scientists, including Newton, Darwin, Einstein, Tesla, Planck and Heisenberg, stopped short of embracing either positivism or atheism.

Modern scientists seldom espouse traditional religions, but that does not make them committed atheists. Positivism is another story. Positivism's and materialism's anti-religious stance is itself dogmatic and must be taken on faith, which entangles its adherents in a contradiction that gives pause to scrupulous scientists. Positivists often find Marxism politically simpatico, though positivists should logically reject the metaphysics in Marxist "dialectical materialism" and doctrinaire political commitments which must be embraced on faith. Islam is the ideology least compatible with positivism and scientism, as well as with science itself.

2. Because the Koran is deemed unalterable truth, and Sharia is valid for all ages and societies Islamic doctrine and practice are highly resistant to change. Both the Koran and Sharia rely on conceptions of time and God that stress stasis, but with an important exception. While the Koran and Sharia are theoretically for all time, the God of the Koran, being absolutely omnipotent, can nonetheless change anything including the past. As Maimonides complained, the Islamic God can even change the rules of logic or geometry or make an elephant the size of a mouse or a mouse as heavy as an elephant. With such a God, the universe becomes like a game in the divine mind, as in the sci-fi movie *The Matrix* where the reality humans believe in is merely a simulation created by a superior mind.

No limits on the arbitrary power of Allah means we can never know for sure what He will or will not do because, so far as we can determine, He ultimately lacks a stable, predictable, or even a discernible nature. Divine unpredictability makes the universe itself unpredictable, which seriously inhibits scientific endeavor and creates a wild metaphysical swing from absolute stasis to the potential for absolute flux. Unpredictability is apparent in Allah's contradictory behavior and unstable temperament manifest in the Koran. Jung might have written an *Answer to Job* for Allah and the Koran.

3. The Islamic and the Christian gods both developed from the substrate of Abrahamic myth, the occident's aboriginal monotheistic mythology. However, they diverged on issues so crucial that their theologies posited different godheads powered by different archetypes. Influenced by ecumenicalism, contemporary Western writers downplay the significance of the ideological and theological bifurcation of Abrahamic myth between Islam and its Western cousins. They fail to recognize that Islam and Judeo-Christianity, taken together, become a manifestation of Jung's two hostile brothers archetype. The opposition of the facets of this bifurcate archetype is essential to its dialectical function. While the brothers may be twins, they still embody two different gods. Jung has shown how the brothers archetype runs through various mythologies, famous pairs being Cain and Abel, Gilgamish and Enkidu, Isaac and Ishmael, Jacob and Esau, Apollo and Dionysus, and Christ and Satan-Lucifer. The later pairing is evident in Jesus's encounter with the devil during his forty days in the wilderness (See *The Unfolding God of Jung and Milton* for my extended treatment of their encounter and relationship.) In contrast to western ecumenical thinkers like Karen Armstrong, Islamic scholars who stress the opposition inherent in a Biblical pair of hostile brothers, Isaac and Ishmael, appear to be closer to Jung's insight into the archetypal dynamics of the Abrahamic tradition.

Myth nourishes the spirit, and theology trains the mind. Modern westerners dismiss theology and theodicy as anachronisms. Though mythology is popular, only a few take it seriously. Their dismissal and neglect has left the contemporary West intellectually and spiritually impoverished and without a solid ethical rudder. The impoverishment contributes to and in part explains the pervasive modern malaise. It also helps explain why so many western intellectuals minimize the darker psychological and political realities of Islam, and downplay the distinctive character and insights of the other great religions. Their lack of a solid ethical rudder is evident in their reluctance to hold Islamists strictly accountable for systematic abuses of women, gays, and Jews. Moreover, the West's desperate turn to obsessive consumerism, in search of an anodyne for lost faith in religion, is a conspicuous consequence of the same spiritual impoverishment and despair which motivates the radical counterturns to Marxism and/or Jihadist Islam among the deeply disaffected.

4. The Abrahamic pairs of Isaac and Ishmael and Jacob and Esau reflect parallel pairs of opposing archetypes, the two hostile brothers and the ego and shadow. Within Christianity itself the split continues as a dialectic embodied in Son of Man-ego versus Satan-shadow. These pairs manifest, for Jung, the dialectical interplay of opposites in the individuation process. Another critical archetypal manifestation is Christ versus anti-Christ which appears in the war of civilizations between Christianity and Islam that began in the life of Mohammad and continued through the Crusades, the conflicts with the Turks, the Spanish Inquisition, and simmers today in the West's struggle with Jihadist terrorism. Applying

Jung's hostile brothers – ego versus shadow archetypal split to Judeo-Christianity and Islam does not make them respectively good and evil. The split manifests a complex dynamic, a dialectic that could yield a synthesis. From today's vantage point such a synthesis seems difficult, distant, and improbable, but the Jungian dialectic of individuation offers reasons to be more sanguine about the longer course of history.

Ideology prefers a war of opposites with each side seeking the destruction of its opponent, while religious motivation accommodates opposites to integrate them in a process that unfolds the civilizational Godhead across time. Ideology leans toward the Manichean. Christian orthodoxy with its Trinitarian God is less Manichean, less fundamentally ideological, and in theory more readily allows transcendence of the warring opposites. It should be noted that inflexible, ego based ideology and its attendant vice, fanaticism, has infected some followers of all three contemporary Abrahamic religions. A notable, example in western society is the species chauvinism that inhibits balanced, respectful debate of the moral issues raised by stem cell research, homosexuality, contraception, and abortion, and by modern man's brutal exploitation of the natural world.

5. Three schools of thought on the nature of God, Whitehead observed, prevailed up until the rise of Islam. Under Roman influence, God was conceived as an imperial ruler in the image of Caesar. The Hebrews saw God as personified moral energy; or Freud might say, God is superego personified. Aristotle sought in one God, instead of the Greek pantheon, the ultimate principle or ground of being, a view similar to that

prevalent in Hinduism. Jung's principles were, Kantian not Platonic, and more in line with Whitehead's process conceptions of time and divinity. Following his studies of hermetic and mystical traditions, Jung saw in God a regulatory entity like the self only on a much vaster scope. That meant God and the self reflect each other, like macrocosm and microcosm. Jung's God is an active, guiding force, but for Jung, as well as for Whitehead, God is not specifically a moral force, not a societal or cosmic superego that protects society by imposing social norms. God as moral nanny is the kind face of God as Caesar, and like Caesar this God embodies ego and superego. Jung's God, reflecting the self, coordinates its subsidiary archetypes such as the ego, the shadow, the anima-animus, the mother-father, and the persona. The self and God seek wholeness, or to use Jung's term the *unio mystic*a, something larger than the objectives of the subsidiary archetypes.

Here Jung's ideas on psychology of religion become more specific than Whitehead's. But Whitehead goes further into religious metaphysics. In Jung completeness and balance, which together create wholeness, are the essential parameters, indeed standards, for deity, self, and personal individuation. Whitehead stresses the love of God for his creatures which motivates the deity to limit His power in order to give others creativity and freedom. Jung saw love more as creative eros than as selfless caritas. He stressed the Quaternity and the mandala as symbols of balance, wholeness, and individuation. The symbol of creativity for Whitehead is not eros, it is written across the night sky in the great spiraling galaxies. Individuation requires positing and resolving, through transcendence of imbalances, the conflicts of opposites. The result is a synthesis achieved through a homeostatic temporal process that Jung termed transformation. The ultimate goal of transformation is balanced wholeness, whereas the ultimate goals for Whitehead are creativity and freedom. The two goals are compatible and mutually reinforcing.

In *Answer to Job* and *Aion*, Jung sees the Judeo-Christian Godhead as an individuation process, set in mythic format, striving for a yet to be achieved wholeness, in short an unfolding God. A religion and a civilization can be judged, he believed, on the wholeness of its conceptions of God, and on the wholeness of its ideals for humanity. What is the good human life? the Greeks asked. They might have also asked, 'what is the good religion, personal code, or political system?' The good strives for and creates wholeness, Jung might have answered, which entails transformation to achieve balance, harmony, and integrity. An essential measure of wholeness, balance, harmony, and integrity is beauty and the creation of beauty, art. Life strives for more than just survival. Like art, life seeks beauty in the forms of wholeness, balance, harmony, and integrity or authenticity.

MEDITATION V:
TIME AND THE ARTS

"The Teleology of the universe is directed toward the production of beauty."

-A.N. Whitehead

1.Art is alive. Life is art. Art can capture the images of life in a timeless moment, as in a painting or sculpture. Art can capture the rhythm, tempo, and processes of life, as in music, dance, poetry, theater, or narratives such as epics and novels. For examples, consider El Greco's "Burial of Count Orgaz," Shakespeare's *Hamlet*, Verdi's *Requiem*, Tchaikovsky's *Swan Lake*, and Tolstoy's *War and Peace*. All display life as an organic process that creates beauty and thus value; for many, including Shakespeare, a key value is wholeness. Art is also a process whereby the values of life become conscious in works of beauty that manifest artistic unity, balance, and harmony to create wholeness. Art is life's offering to The God in Time.

2. Like art, life, including individual lives, must have limits, a beginning and an end, a frame to the picture. Otherwise it lacks form, direction, and purpose: without limits it becomes like elevator music. Life is art, and art is alive. The plans of life are art effected. Nietzsche observed

that in making the conscious decisions that shape our lives we resemble artists constructing their narratives: thus, we can judge a life as we might judge a literary narrative. Man's ability to make art consciously separates him in degree, though not in kind, from animals and plants that make their art instinctively and whose beauty is unconscious. That time, like art, is defined through reflection on our feelings about and perceptions of change would seem a truism. Thinking about time and making art have the same origins in the human drive to live life more consciously and more fully.

3. Man is a social animal. His intelligence is partially instinctual, but not largely instinctual, as is more the case with other brainy creatures, like dolphins, apes, cats, or crows. Man's intelligence appears to develop mainly in a social and cultural context finding a hospitable bubble in time. Culture, especially technology and above all language, defines, extends, and develops man's native intelligence. Everyman perceives and understands time consciously within and through a cultural context. Heidegger refers to each man's specific cultural context, their bubble in time, as his historicity.

Science, technology, religion, and philosophy are mainstays of advanced cultural contexts, along with, literature, and the arts. Magnificent artistic creations can better convey deeper truths than most writings expounding religions or philosophies. Their historicity, the peculiarities of their time bubble, gives their truths a specificity more abstract works lack. Joyce's *Ulysses* and Tolstoy's *War and Peace* are renowned works that display high universality and very specific historicity. We learn a great deal from them about life in Dublin in June

1904 and in Moscow in the Napoleonic era. Paradigm shifting science, such as Newton, Darwin, and Einstein, may be the most universal cultural artifacts: it often carries momentous implications for our conceptions about time and our world.

Technology is oriented toward specific tasks, the measurement of time being one of its most important tasks. Where technology usurps or dominates the arts, it creates a malaise that can only be dispelled by freeing the arts. The arts create and express values, technology implements the purposes values express. So technology must ultimately be subordinate to the arts which are the primary source and expression of values in either a healthy culture or a well individuated psyche.

4. The similitude of art and life can tell us something about the major religions and their images of "God." One might expect the religion that best captures, and whose art most fully represents, the entire range of human life to be the one with the most balanced and comprehensive, indeed the truest, understanding of God. Shakespeare is the most celebrated poet and dramatist of all time. His *oeuvre* is profoundly Christian, though not dogmatically or even predictably so. His felt understanding reaches beyond the grasp of mere theologians and philosophers to the ethical and psychological heart of Christianity, and in *King Lear* it lays bare the archetypal depths and mythic core of the Judeo-Christian Godhead. More consistently Christian writers, like Dante, Milton and Tolstoy, are among the handful of the greatest writers, as are less consistent Christians like Cervantes, Dostoevsky, and Joyce. Bach is deeply Christian; as are some of Mozart, Beethoven, Verdi's and even

Wagner's triumphs. The art of many of the world's finest painters has often been expressly Christian: for example, Giotto, Botticelli, Raphael, Titian, Michelangelo, Bosch, Durer, El Greco, and Velasquez. The superior achievements of Christian generated art and culture are a plausible argument for Christianity over rival religions. The contention being, that Christian culture's total image of God, the Christian Godhead and its constituent archetypes, are a truer representation of the full human psyche, and, therefore, its artists were able to represent more fully and truthfully the pageant of human life.

Islam's implicit downsizing of the Godhead to a single archetype has fostered prohibitions that inhibit development of painting, sculpture, literature, and music. Since representation of images is forbidden, Islamic art focuses on fixed geometric patterns for rugs and ceramics, and on architecture, areas where its achievements have been glorious. Architecture is a powerful but static, and therefore more constrained, art form. It strives for timelessness, as do the geometric patterns of rugs and ceramics. Yet even the pyramids age, and vast stone monuments are ground to sand leaving, as Keats tells us, only Ozymandias's pretentious fragments. Architecture is the marriage of geology to geometry. Painting is closer to chemistry, and in great painters it becomes a form of alchemy revealing the self in quest of the high alchemical goals. Many believers think that Islam forbids use of musical instruments, and so Islamic music is often limited to vocalized prayer and devotion. There is no Bach, Mozart, or Verdi in Islam. Likewise, there is no Islamic Michelangelo, Velasquez, or Bernini. Dogmatism obstructs and stifles literature, and so there is no Islamic Cervantes, Shakespeare, Dostoevsky, or James Joyce.

Nor do we find artists comparable to these in the Hindu, Buddhist, or Confucian east. These great western artists are a few examples, out of a multitude, of how conceptions of time and Godhead are based on psychological assumptions, predispositions, and yes archetypes, that exert profound influence throughout a culture.

5. Time is perceived thru changing sensations, though intuition gives us the sense of duration. But our apprehension of time varies with and is modulated by feeling. The first clocks we encounter are the rhythms of heartbeat and respiration. Rhythm is our perception of groupings and patterns in the flow of time. The heartbeat measures the flow of time in meter, as does the music of our feelings. We modify all experience by coordinating it with the variations in the rhythms of our feelings within the flow of time. Music expresses the interdependence of time and feeling. Time and feeling are interdependent for men and other sentient creatures. Animals express feeling directly with minimal modulation from thought; humans are more conscious of their feelings. We infuse feeling with intuition and shape, curb, heighten, or direct it by thought.

Time and feeling are the form and substance of value, there is no value without time shaping feeling and feeling giving substance to time. Nietzsche said that values arise from action; true, but values also arise from joy, suffering, work, and reflection in solitude—that is, being alone in and with time, which Whitehead saw as essential for religious experience. Jung stressed how values arise from encounters with the self that create mental states whose unusual intensity seems to place them outside of time. Common examples of feelings so intense that consciousness momentarily transcends time are the ecstasy of intercourse, mystical states, or even states of panic. Such peak experiences can change the course of a life.

6. A universe without time would be without feeling or value, it would be static, lifeless. Indeed, life, like the universe itself, depends on time, exists only in time, and all life forms have their own, distinct adaptions to time which are crucial definers of their identities and their historicity. A hypothetical universe where all is eternally "perfect" would be timeless. To envision what the quality of timelessness might be like, try to imagine not silence but a vast, deadening monotone devoid of all variation and music. Shakespeare's words might apply to those philosophers that exalt timeless, eternal "perfection" over the music of time:

The man that hath no music in himself,
Nor is not moved with concord of sweet sounds,
Is fit for treasons, stratagems, and spoils.
The motions of his spirit are dull as night,
And his affections dark as Erebus.
Let no such man be trusted. Mark the music.
Merchant of Venice V.i

Contrariwise, like the Pythagoreans, Shakespeare believed the harmonies of music can help us maintain sanity, and restore balance and harmony to the soul. Indeed, all the arts have healing qualities for the performers, their audiences, and even society; for example, catharsis in tragedy. Jung utilized the healing power of painting for both himself and his patients.

7. Mark the music. Music is the language of pure feeling, expressing, as feeling always must, the rhythms and cadence of the soul caught in time. Music is, as Gotthold Lessing observed, the art of time, and it subsists wholly within time. Unlike paintings, sculptures, and architecture, musical compositions lack instantaneous appeal, they must be experienced across time. For those who think time an illusion, listen to the music. Music is the alchemy of time, it projects feeling into sound by differentiating sound into durational patterns in time, such as rhythm, melody, and tempo. All music has "tempo," the very word derives from words for time. Tempo modulates the emotional effect of rhythms, so that faster is active and cheerful, slower is contemplative or somber. Music cannot exist without rhythm. We might speculate that feelings cannot exist without rhythms. They do not exist without a heartbeat! Neither music nor feeling can exist apart from time.

No feeling remains the same for long, all undergo modulation with time. Likewise, in music the sound must continually change with the tone, rhythm, meter, melody, etc. Musical structures are rhythmic, musical form is metric. Rhythm depends on changes in pitch, intensity, harmonic function, and timbre, all of which directly express and describe feeling and modulate change across time. Duration sets the temporal relationship between one tone and the next, and accents color the tones. Motion in music is primarily governed by changes in intensity and duration. Melody is a varied sequence of notes that express meaningful shifts in feeling. Melody and harmony are set by the rhythmic progression of time. Contrapuntal melodies can express complex, interactive feelings or temporal situations. All of these factors working together give music the power to represent in real time complicated, layered emotions along with dramatic and simple emotions.

In a motion picture the visual tract carries narrative, but the music sound tract expresses actions and shapes the feelings of the characters; it guides the responses of the audience, and punctuates visuals and dialogue. Movies almost always have musical sound tracts, because their message is best carried when parallel levels of sight and sound, thought and feeling, are activated and harmonized to reinforce each other. The music makes it possible for the film to impact our entire being. Its multifaceted impact explains much of the power and popularity of cinema. Thought may appear to be outside time to Plato and other stasis philosophers, but music imposes a major limitation on the scope of their theories by demonstrating how feeling parallels thought and how both are integral with time.

Dance, a close partner to music, adds to music a spatial dimension. Sculpture and architecture subsist in space. Painting subsists in the field of vision, but also in space. Painting is a kind of alchemy that involves actual chemistry in mixing colors, tones, shadings, and shapes to apprehend feelings and thoughts in a physical representation all can see. In painting color and shading convey feeling, line and shape convey thought.

Poetry and literature require a broader involvement of imagination than the visual and aural arts where direct response comes easily. All good poetry is pictorial and musical. Language when spoken or sung poetically can become a form of music: just read aloud any poem by T. S. Eliot. But there is more to poetry than its music. Poetry depends equally on thinking, feeling, and imaginative deployment of the senses in its attempt to integrate them through artistic intuition.

Drama combines poetry, music, vision, spacing, and timing. Without superb timing drama fails for lack of tension and suspense, i.e. lack of drama. Timing is essential in all comedy; wit, jokes, and humor fall flat if not well timed. Great comedians have great timing. In cinema the script is usually less poetic than in drama, visuals have a larger role, and the sound track is more critical. (Shakespeare would have loved the sound tracts of some of our best movies.) Timing is as vital in cinema as in drama. Both are multifaceted art forms that create transient pictures and even sculptures fleeting through time, especially when the movement of the players takes on elements of dance, as can happen in well-choreographed action movies. All the kinetic arts seek exposition and integration of the range of human faculties across time:

> *O chestnut tree, great rooted blossomer,*
> *Are you the leaf, the blossom or the bole?*
> *O body swayed to music, O brightening glance,*
> *How can we know the dancer from the dance?*
> W.B. Yeats, *Among School Children*

Yet music remains the most completely temporal of the arts, which allows it to become the purest expression of feeling. Some say music is pure feeling without the encumbrance of language, though language is often musical on its own. Music is the voice of the soul and the sound of the spirit moving through time. As Yeats knew, it is the soul that sings and its music can enliven even the aged with new meaning:

An aged man is but a paltry thing,
A tattered coat upon a stick, unless
Soul clap its hands and sing, and louder sing
For every tatter in its mortal dress,
Sailing to Byzantium

Music can be a glorious avenue to deepening experience of Time, God, value, and the meanings of life. Just listen to Bach, Beethoven, Mozart, Wagner, or even the Beatles. The true musician is the supreme alchemist of the soul who transforms through imagination the dross of our raw, chaotic feelings into the gold of art and felt insight.

8. Imagination is the peculiar faculty of the artist. It allows great talents like Shakespeare immediate, intuitive insight into emotional truths, direct apprehension of beauty, and glorious representations of the power, motion, and structures of time. The artist's imagination can create out of pure feeling something tangible that can be sensed and heard physically and understood intuitively, if not always intellectually. It transforms what is individual into universal art that many can apprehend and enjoy. Music especially spans and unifies the psychic and the material realms as the alchemists sought to do. The alchemists' unconscious motivating goal, Jung believed, was individuation or full realization of the self. The closely related goal of musical alchemy is integration of body and soul. Music, in expressing the soul, can soothe the body and act as a healer of souls.

Is the universe itself God's art making, an exercise of divine imagination within the medium of time? Whitehead said, "He (God) does not create the world, he saves it; or more accurately, He is the poet of the world, with tender patience leading it by his vision of truth, beauty, and goodness." So it makes sense to look to the greatest of our poets to see how God saves the world by His, and the poet's, vision of truth, beauty, and goodness working through time.

MEDITATION VI:
TIME, THE GOD OF SHAKESPEARE'S WORLD

"What does it mean to be human? —perhaps the most important question, if you happen to be human. My leopard cat Ariel, who has studied me assiduously for fourteen years, also appears to believe in it's important. Indeed, as mankind's impact on nature has grown over the last five centuries the question's importance resounds across the biosphere. The answers should also give us indications as to what the optimal God for all humanity might be like. More and far better than any other body of writing, the opus of William Shakespeare answers the question in stories and poetry that dramatize the nature, glories, failings, and meanings of humanity. A defining feature of the human condition in Shakespeare is the individual caught in time and challenged by time. Shakespeare does not assert that time is God so much as he demonstrates it dramatically and poetically. In Shakespeare time assumes the role other writers assign to God.

Great writers are like the Biblical prophets or Homer in that they create what Harold Bloom termed the "canon" of a civilization which embodies that civilization's wisdom. The creation of the Western canon did not stop with the Old Testament prophets, Homer, or the New

Testament, but continues into our age. Canon writers often consciously (and unconsciously) follow primitive myths such as those found in Homer and the early Greeks, and in the Bible: James Joyce's *Ulysses* is a case in point. The canon becomes a continuing revelation of man's interactions with the divine and a record of a civilization's different views of the divine. Plato and Aristotle, who saw poets as speaking for the gods, validated this standard for posterity. Shakespeare may be a prophet for all humanity, more than Homer or Dante, even more than Jesus and Buddha. Shakespeare's preeminence was recognized by critics and by popular acclaim within a century of his death. But its implications remain for future generations to develop and digest fully. Remember, it took a long time for the Bible to achieve the popular status and influence it has enjoyed since Gutenberg.

Shakespeare in his study

Shakespeare came one and a half millennia after Jesus, the most influential of the purely religious prophets. Mohammad being an ideologue, as well as a military and political genius, was a prophet more in the qualified sense of Marx. Shakespeare has a perspective different from, broader, more profound, and more universal than any of the prophets. From Jesus forward, the true prophets, the bringers of new revelations for humanity, tended to be seers or poets leaving a specific body of literary work, rather than shadowy figures couched in myth such as Moses, Homer, Buddha, Zoroaster, or Jesus himself. Besides Shakespeare, high among the cardinal seers of the Christian era and its canon would be St. Augustine, Dante, Cervantes. and Milton. The last two hundred years offer new seers, literary and otherwise, such as Melville, Dostoevsky, Nietzsche, Bergson, Whitehead, James Joyce, and Carl Jung. These moderns were directly challenged by and forced to think and respond in new ways to the transformational social upheavals, religious doubts, and widespread spiritual malaise of the modern era. The most challenging breakthrough religious insights among the latter day seers may be from Nietzsche, Whitehead, and Jung. But it is too early to fully evaluate their impact. Time will tell.

All of these great seers deal with time, implicitly and sometimes explicitly. Shakespeare, being in a rank by himself, has the foremost claim upon our attention. His vision of time, touching nearly all important aspects of the human experience, is not to be found in any one play, sonnet, character, or passage. It unfolds through the paramount literary pageant, his opus, even as time itself unfolds in human history, earthly life, and the universe. Like life, Shakespeare's entire opus shows us how through suffering, joy, terror, laughter, folly, heroism, villainy, exaltation, insight, and emptiness we create truth, beauty, and value, and unfold meaning in time. Let us, then, consider Shakespeare's essential passages on time.

King Lear, the supreme literary tragedy, presents the crucial insight into the ultimate mystery of life and of universe--time itself. In perhaps the most profound reflection on time in Shakespeare or anywhere, the nature of time is at once excruciatingly painful, powerful, beautiful, and simple:

> *And my poor fool is hang'd! No, no, no, life!*
> *Why should a dog, a horse, a rat, have life,*
> *And thou no breath at all? Thou'lt come no more,*
> *Never, never, never, never, never!*
> King Lear. V.iii

The prime existential feature of time, *King Lear* shows, is its implacable irreversibility which makes every event, and every life unique and thereby gives value to all. Time is the great value creator. Tragedy reveals time the creator and time the destroyer locked in agonized conflict. For conflict is essential to tragedy, to time, and to the creation of reality itself which according to modern physicists began with an incomprehensibly vast and powerful explosion. *King Lear* also reveals, as no other work, how God, as the Christian Godhead, unfolds in the archetypal stages Lear passes through. As I have explained in my book *Identity in Shakespearean Drama*, these are the Yahweh, Job, Prometheus and Cordelia stages.

Critics note parallels between Lear the patriarchal tyrant and the Old Testament Yahweh, as well as between Lear in the storm and Job. They stress Cordelia as a Christ figure, and note Edmund and Edgar as the hostile brother's archetype. The hostile sisters archetype, Goneril and

Regan against Cordelia, is less explored. In myths and fairy tales two evil sisters often oppose a single good sister. Together the bad sisters represent the dark or shadow *anima*, and Cordelia represents the good anima or wisewoman archetype. Yet the archetypal schema of *King Lear* remains occluded because neither the dynamism of Prometheus nor the parallels to Job nor the wiseperson role of truthful Cordelia is recognized. Critics search in vain for realistic motivations in characters that are essentially mythic figures. They notice that Cordelia is Christlike, but not that Lear is the crucified one. Lear tells Cordelia upon their reunion:

> *Lear*, Thou art a soul in bliss, but I am bound
> Upon a wheel of fire, that mine own tears
> Do scald like molten lead. IV, vii
> *Cordelia*, "Sir do you know me
> IV, vii

Lear is not alone in not knowing who Cordelia is, none of the critics seem to grasp that as a mythic figure she represents the Paraclete, the *anima* of the Godhead embodying truth and love and therein the wiseperson archetype.

Colm Feore as King Lear and Sara Farb as Cordelia in King Lear Photo by David Hou.

The iconography of the final scene of *King Lear* is a reversal, almost to the point of parody, of Michelangelo's Pieta. In place of the sublime sadness of Mary, mother of God, gently holding her slain Son, we get the Promethean rage of the father grasping his slain daughter, crying, "She's gone forever. /I know when one is dead and when one lives;/ She's dead as earth." It's as if Shakespeare presents a passion play with the sexes reversed, along with the moral of his tale. Instead of redemption from all sorrows in the promised end of resurrection, we get the image of that horror, the implacably power of irreversible time. One element remains unchanged, Jesus and Cordelia both embody the prime values of love and truth.

King Lear, the most profoundly tragic of all literary tragedies, needs to be studied alongside and in contrast to the *Book of Job* and the Gospels. The play is as crucial as the *Book of Job* and the Gospels to understanding the unifying and directing archetypes of the Godhead that are constellated within and across the collective unconscious of Judeo-Christian Western culture and civilization. In his *Answer to Job*, Jung delineates the first two stages of the revelation of Western deity, the Yahweh and Job-Christ stages. He missed or misunderstood the next stage, Promethean repudiation of dogma for existential truth, whose crucial emergence in the Renaissance powered the development of science and technology. Jung, interested more in alchemical writings than in Renaissance literature, left it for posterity to detail how specific works of literature constellate crucial archetypes in the history of civilizations the way signal events, important dreams, and synchronicities do in the lives of individuals. He never systematically analyzed Western Civilization as he might a patient. However, Jung appreciated the power of the unique insight given to artists, poets, and seers, and might well have conceded that the greatest of artists, Shakespeare, could have intuitively reached further and deeper into the archetypes of the Godhead than others had.

Here are my previous remarks on the Promethean archetype that Jung overlooked, but which assumes crucial importance in *King Lear* and in the development of the Christian Godhead, the Western psyche, science and technology, and our cultural collective unconscious:

The Promethean archetype dominates the final stage of Lear's quest for consciousness through his confrontation with injustice, death, evil, and ultimately God. The manifestation of the Promethean archetype in King Lear bears important similarities to its form in Greek myth. And Lear's cry, "Oh Regan, she hath tied/Sharp tooth'd unkindness, like a vulture here." (II.iv) may indicate that Shakespeare used Promethean analogy deliberately. While in transit from the Yahweh to the Job archetype, Lear displays something like a Promethean stance both in denouncing social and metaphysical evils and in showing compassion for fellow sufferers. Prometheus name means forethought and he has often symbolized consciousness. . . . Lear in his last scene exhibits Prometheus's insight into cosmic justice, his sense of divine betrayal and his defiant assertion of the dignity of human life. But the Promethean archetype derives its peculiar significance and impact from the largely unconscious way Shakespeare the artist relates it to the Job, Yahweh, and Christ archetypes to form a comprehensive vision of (the Judeo-Christian) God. . . . Orthodox Christianity's characteristic emphasis on the Job stage and the vindication of Yahweh and reduction of evil to privatio boni inhibited movement beyond this stage to the unflinching recognition of evil and unfettered protest of true Promethean consciousness. In King Lear evil gains recognition as a far more powerful and fundamental force than mere privatio boni which, Jung suggests, denies dignity to tragic passion by rendering it and the evils that cause it ultimately unreal. The Promethean stage compensates for and transcends the Job stage because it fully recognizes the reality (and power) of evil.

Full recognition of the reality of evil entails full recognition of the irreversibility of time, of time's power over human life. Tragedy is the supreme literary art form of Western civilization whose development of tragedy far exceeds that in any other civilization. Tragedy's supremacy, and *King Lear's* supremacy within the genre, derives from its revelations of God, the human condition, and, ultimately, its unflinching recognition of time's implacable irreversibility. The Promethean archetype provides the key to that recognition in *King Lear*, and the key to the Godhead archetypes that silently powered our civilization to become pre-eminent not only through science and technology but in the arts and philosophy. The Promethean archetype brings an unflinching recognition of truth which Edgar proclaims in the closing lines of King Lear that also serve as Lear's epitaph:

The weight of this sad time we must obey;
Speak what we feel, not what we ought to say,
The oldest hath borne the most, we that are young
Shall never see so much nor live so long.
Exeunt with a dead march. V, iii

To cite my comment: "Can there be any doubt that Shakespeare would have us speak what we feel and not what we ought to say, by recognizing the uniqueness in pain and veracity of the vision of *King Lear?* It is a vision of unalloyed evil, unspeakable torment, and irredeemable loss, of undaunted love and the dignity and wholeness that truth and love can give. Following the archetypes, it is a vision of God. To know and to feel so much as we can in *King Lear* brings us as close as poetry can to seeing the faces of God."

Joseph Marcell as King Lear, Globe Theater, Annenberg Center

Harold Bloom appeared to have sensed the importance of *King Lea*r in the development and manifestation of the western Godhead when he noted that after Lear God the Father was never again a sympathetic figure in English literature. The culmination of Shakespeare's profound, but hard to grasp and even harder to accept, insights about the Father figure in the Godhead came centuries later in Jung's analysis of Yahweh in *Answer to Job.* It is a true breakthrough work whose insights have yet to be absorbed by theology and philosophy. The questions Jung raised through the entirety of his work on theodicy and the archetypes of the Godhead form an island of light in the dusky archipelago of modern theological gloaming.

The opposite and complement of *King Lear's* insights on time and God is found in the unremitting despair of *Macbeth.* Macbeth and his Lady are souls that self destruct. They commit spiritual suicide through worshipping power and thereby rejecting love and truth as guides to life. For all his tragic faults, Lear loves truly and the good characters in the play, Lear, Cordelia, Kent, Edgar, and the Fool, are all capable of loving with their entire beings. These good characters are avatars of truth ready to sacrifice their lives in defense of uncompromised truth. Their unfailing courage in the cause of truth is the ultimate proof of their love.

Macbeth and his Lady, choosing power over love, advance their ambitions by lies and self- deception. Power attained by lies gains them nothing of value. Macbeth ends his life in the absolute emptiness of utter despair. Lady Macbeth appears to embrace her lord's final cynicism from the beginning:

Things without remedy
Should be without regard. What's done is done. III.ii

Yet in the end she is less able than Macbeth to bear the ravages of the guilt she at first so facilely dismisses. She takes to sleepwalking with a candle muttering about the irreversible, evil deeds she and Macbeth have done and the indelible guilt she has incurred:

Here's *the smell of blood still. All the perfumes of Arabia will not Sweeten this little hand.* V.i.

Choosing ambition and power over love and truth is the root of Macbeth and Lady Macbeth's tragedy: indelible guilt and relentless despair are the consequences. Stronger, ultimately, and more articulate than his prosaic Lady, Macbeth's mind and sanity remain clear and steady in the face of utter defeat and irreparable loss. Macbeth is given western literature's most memorable outcry of despair at time's destructive power and the meaninglessness of a life lived without truth or love:

Tomorrow, and tomorrow, and tomorrow,
Creeps in this petty pace from day to day
To the last syllable of recorded time;
And all our yesterdays have lighted fools
The way to dusty death. Out, out brief candle!
Life's but a walking shadow, a poor player,
That struts and frets his hour upon the stage
And then is heard no more. It is a tale
Told by an idiot, full of sound and fury
Signifying nothing. V.v.

Powerful as it is, Macbeth's despair is far from being Shakespeare's last word on time and human life. It is only one perspective among many which taken together constitute a range of experiences of time unsurpassed in depth and comprehensiveness.

Macbeth's final despair arises out of his strategy toward time which is tragically limited. Macbeth and Lady Macbeth are both obsessed with the future and ever eager to trade present security, peace, and happiness for future advancement. Lady Macbeth, especially, is never satisfied with the present but, always looking for a better day tomorrow:

These letters have transported me beyond
This ignorant present, and I feel now,
The future in the instant. I.v.

For her the present is only a means toward a better future, all value lies in the future. Tragically, the present has no other value for those ruled by raw ambition. The witches reveal the future, and Macbeth's fascination with them is a fascination with the future. Just as the past is the realm of memory, the future is the realm of the imagination, and Macbeth, its denizen, is wildly imaginative, Like Milton's Satan, Macbeth and his Lady are victims of ambition inflated by diseased imagination. Obsessed with their future, they can find no happiness in the present and no value in or guidance from the past. Their tragic dilemma becomes manifest when time and the future close in on them leaving their past a tale told by an idiot signifying nothing.

An earlier Shakespearean king who prefigures Macbeth, Richard II, learns a more prosaic and practical lesson about time and his own tragedy: *I wasted time, and now doth time waste me.* (Richard II, V, V, 49) The need to use time wisely, to be a good steward, is a recurrent theme throughout Shakespeare's plays and in his sonnets. It lies at the core of the bard's ethical and prudential views. It is a specifically Protestant conception that reflects the influence of the Reformation on Shakespeare and signals a value associated with Prometheus the fire bringer, the ideal of progress. Judicious management of time would become a driving force behind the industrial revolution and the rise of modern technology.

Wasting time is one of Falstaff's pre-eminent flaws, but the result in his case is comedy.

Now, Hal, what time of day is it lad? Falstaff,

What a devil hast thou to do with the time of the
day? Unless hours were cups of sack and minutes
capons and clocks the tongues of bawds and dials the
signs of leaping-houses and the blessed sun himself
a fair hot wench in flame-coloured taffeta, I see no
reason why thou shouldst be so superfluous to demand
the time of the day.
Henry IV Part I, II.

Hal's retort shows that in an untroubled comic world time's irreversibility does not seem to affect us, at least we can ignore it. Falstaff, almost Macbeth's opposite, lives entirely in the present with little regard for the future. Falstaff resides in a comedic timeframe, but the comedy wanes as he ages. *I am old, I am old*, he declares. Falstaff displays throughout a comedic view of mankind's fate, prime limitation and dilemma, time. But in Henry IV Part II time the destroyer, always looming, reasserts its power. *I care not; a man can die but once; we owe God a death.* III,ii. declares Falstaff. Falstaff lives and maintains his identity within the Valhalla of his comedic timeframe. Although the Falstaffian timeframe finally crumples, it does not disappear. It is transformed by art, its transformation manifest in Falstaff's own passage to the great beyond:

> *Nay, sure, he's not in hell: he's in Arthur's*
> *bosom, if ever man went to Arthur's bosom. A' made*
> *a finer end and went away an it had been any*
> *christom child; a' parted even just between twelve*
> *and one, even at the turning o' the tide: for after*
> *I saw him fumble with the sheets and play with*
> *flowers and smile upon his fingers' ends, I knew*
> *there was but one way; for his nose was as sharp as*
> *a pen, and a' babbled of green fields. 'How now,*
> *sir John!' quoth I 'what, man! be o' good*
> *cheer.' So a' cried out 'God, God, God!' three or*
> *four times. Now I, to comfort him, bid him a'*
> *should not think of God; I hoped there was no need*
> *to trouble himself with any such thoughts yet. So*
> *a' bade me lay more clothes on his feet: I put my*
> *hand into the bed and felt them, and they were as*
> *cold as any stone; then I felt to his knees, and*
> *they were as cold as any stone, and so upward and*
> *upward, and all was as cold as any stone.*
> Henry IV Part II, II, iii

This sublime account of Falstaff's death from the mouth of a bawd showcases art's power against death and time the destroyer, one of the great Shakespearean themes and one of his conscious designs as an artist from the Sonnets forward. The power of art over death is ultimately the power of art to heighten and enrich human experience and consciousness.

If Macbeth, Hamlet, Othello, and Lear are the great tragic accounts of death, Falstaff's stands alone as the most magnificent comic account.

While Falstaff appears to be time's victim, Prince Hal seeks to be its master, "*I'll so offend to make offense a skill, Redeeming time when men think least I will.*" Henry IV Part I, I, ii. But we know that the hero can master time for only so long, in the end time's omnipotent irreversibility masters all. Hal reigned as Henry V for a brief 9 years. His death at age 36 initiated a long reign of chaos during the Hundred Years War, as Shakespeare's audiences would have known. "*We are time's subjects, and time bids be gone.*" Henry IV I, iii. Hastings reminds them, in a sober, humdrum but accurate statement of the insistent limitations time imposes on each life, down to its individual moments.

Shakespeare's Sonnets counter this drab view by stressing the creativity of time through art, and demonstrating the ability of art to turn time upon itself. In the Sonnets time is Janus faced: it is both the great destroyer and the creator of possibilities for making art that transcends time the destroyer.

Like as the waves make towards the pebbled shore,
So do our minutes hasten to their end,
Each changing place with that which goes before,
In sequent toil all forwards do contend.
Nativity once in the main of light,
Crawls to maturity, wherewith being crowned,
Crooked eclipses 'gainst his glory fight,
And Time that gave, doth now his gift confound.
Time doth transfix the flourish set on youth,
And delves the parallels in beauty's brow,

Feeds on the rarities of nature's truth,
And nothing stands but for his scythe to mow.
And yet to times in hope, my verse shall stand
Praising thy worth, despite his cruel hand.
Sonnet 60

Sonnet 60, possibly the most profound of Shakespeare's sonnets, is universally acknowledged to be one of the world's greatest poems. Time creates all and destroys all that it creates, we are not even the same individual we were yesterday. Yet Shakespeare raises the hope that art can transcend and thereby defeat time the destroyer, that the triumphant consciousness of the true artist can harness time the creator to fashion artifacts that express abiding truths and beauty. Jung might say that Shakespeare's poetry endures because it so magnificently portrays the archetypes that shape the range of human experience, overshadow every era, and characterize any specific time period. Shakespeare's view of the irresistible power of time seems irreligious because it appears to assume time prevails over the ego-like Gods of parochial religion. Nevertheless, art, Shakespeare hopes and believes, can be a match for time the destroyer and may offer enduring values which can light our most reliable paths to spiritual Truth and even to God.

In Sonnet 60 there is no traditional God, no immortal soul etc. Time Itself is God and the artist is Time's prophet. The sonnet ends with a paradox in that art, Time's greatest creation, may be Time's own self-transcendence. Through art, time spirals forth to create something novel rather than getting stuck reiterating a Viconian circle or riding the arrow of time into ineludible destruction. Time's spiraling quality is key to its peculiar artistry and divine creativity, as Whitehead shows. Here and elsewhere, Time the creative artist, and the ultimate destroyer, emerges as the veritable God of the world of Shakespeare's plays.

In Shakespeare love is an art and there is art in love. Just as creative art can defeat destructive time so can creative and procreative love defeat time the destroyer. The destructive forces of time, the Bard firmly attests, will in the end prove impotent against time's creative power through art and through love:

Let me not to the marriage of true minds
Admit impediments. Love is not love
Which alters when it alteration finds
Or bends with the remover to remove
O, no! it is an ever fixed mark,
That looks on tempests and is never shaken;
Whose worth's unknown, although his height be taken
Love's not Time's fool, though rosy lips and cheeks
Within his bending sickle comps come;
Love alters not with his brief hours and weeks
But bears it out even to the edge of doom.
If this be error, and upon me prov'd
I never writ, nor no man ever loved.
Sonnet 116

Indeed, love is the greatest force for good in Shakespeare, the force that embodies creative time, and defeats destructive time by transcending it. True love is not Time's fool, it alters not with hours or weeks, love resides in the eternal present. Macbeth and his Lady were Time's most tragic fools relying exclusively on future gains to give their lives meaning and joy. Live not in hope for a better future, the wise Bard counsels. The best, the most enduring and rewarding good, we can achieve in life is to love and create now, in the present moment.

As You Like It presents a range of comic perspectives on time. It also contains the most extended discussions of time in Shakespeare, and compelling dramatizations of the value of loving in the present moment. One of the bard's most oft quoted speeches is Jacques's lofty and detached review of the ages of man. Readers sometimes forget that Jacques is just a character speaking, like Macbeth and all the other personae in Shakespeare's dramas. His limited viewpoint is no more Shakespeare's own than Macbeth's "sound and fury" speech is the bard's last word on the meaning of life.

All the world's a stage,
And all the men and women merely players:
They have their exits and their entrances;
And one man in his time plays many parts,
His acts being seven ages. At first the infant,
Mewling and puking in the nurse's arms.
And then the whining school-boy, with his satchel
And shining morning face, creeping like snail
Unwillingly to school. And then the lover,
Sighing like furnace, with a woeful ballad
Made to his mistress' eyebrow. Then a soldier,
Full of strange oaths and bearded like the pard,
Jealous in honour, sudden and quick in quarrel,
Seeking the bubble reputation
Even in the cannon's mouth. And then the justice,
In fair round belly with good capon lined,
With eyes severe and beard of formal cut,

Full of wise saws and modern instances;
And so he plays his part. The sixth age shifts
Into the lean and slipper'd pantaloon,
With spectacles on nose and pouch on side,
His youthful hose, well saved, a world too wide
For his shrunk shank; and his big manly voice,
Turning again toward childish treble, pipes
And whistles in his sound. Last scene of all,
That ends this strange eventful history,
Is second childishness and mere oblivion,
Sans teeth, sans eyes, sans taste, sans everything.
As You Like It. II, vii

The seven ages of man, set forth in glittering stereotypes, are timeframes that function like acts in the drama of a generic individual's life. Being a dramatist, Shakespeare laid out his stories on a framework of smaller timeframes, or scenes, within the larger acts all framed by the play itself. The act-scene structure of Elizabethan dramas provided a model for his thinking on time, a model that is widely useful in the world outside the theater. The Seven Ages details man's physical and social journey through life in a set of minatures that are at once uniquely memorable and recognizable by all. It is one of Shakespeare's key statements on our ways of structuring time. Accessible to all, it does not need detailed explication, except to note, first, that it applies to the common, generic person, not to the extraordinary, not to Hamlet, Falstaff, or Lear, or to Socrates or Jesus. Second, it omits lived time as process, or historicity, and strives to follow a static structure like illustrations for a calendar.

All the world's a stage and men are but players, is a key phrase for understanding time in Shakespeare: it links the lives of men to dramatic artistry, and makes time the medium of dramatic art. Indeed, for Shakespeare time is the prime medium of the arts that most engaged him, poetry, drama, and music. These arts provide ready and effective means of transcending time into the realm of universal archetypes, as we can see in Sonnet 60. It is evident throughout the plays and the sonnets that for Shakespeare art can further psychological individuation by transcending present time to create a transformative metastance upon the entire temporal process. Indeed, genuine art, whether in its creation or in the viewer's full experience thereof, partakes of personal individuation, and each instance of creation or viewing can become an exercise that furthers wholeness.

Place can profoundly affect time, as the characters in As You Like It learn. Or time can create its own places, like an island in time, as with holidays, a vacation or, on the lam in a region of outlawry where clock time doesn't seem to matter. The forest of Arden is all of these:

ROSALIND

I pray you, what is't o'clock?

ORLANDO

You should ask me what time o' day: there's no clock
in the forest.

The forest of Arden begins as a timeless place; but when town and court sophisticates invade it, they bring with them time's woes and pressing demands, thereby altering the place almost as much as a buzz saw or an invading army might. Indeed, specific places and their times cannot be separated, except in the mind and there only temporarily.

The activity and place of the perceiver of time, Rosalind maintains in an oddly Einsteinian deja vue, is relative to and sets the speed of time.

> ***ROSALIND***
>
> *Then there is no true lover in the forest; else*
> *sighing every minute and groaning every hour would*
> *detect the lazy foot of Time as well as a clock.*
>
> ***ORLANDO***
>
> *And why not the swift foot of Time? had not that*
> *been as proper?*
>
> ***ROSALIND***
>
> *By no means, sir: Time travels in divers paces with*
> *divers persons. I'll tell you who Time ambles*
> *withal, who Time trots withal, who Time gallops*
> *withal and who he stands still withal.*

Not only does time vary with the place, and shape the place, individuals vary with the time and their variations in return shape their perceptions of the place. A specific time is like a picture with a figure or figures who give meaning to and perspective upon their context, the place, and vice versa.

Thus, the pace of time alters with the person, his/her situation, and activity. For Rosalind and Orlando, time is more dynamic than for the detached, sardonic Jacques. For him time is regimented like a mechanical clock or a calendar, for the two young lovers time is more like a river flowing. In their cases two rivers flowing toward the place where they joyfully unite as one. Like a river, time can move turbulently through its rapids, or amble lazily through a sleepy delta.

ORLANDO
I prithee, who doth he trot withal?

ROSALIND
Marry, he trots hard with a young maid between the
contract of her marriage and the day it is
solemnized: if the interim be but a se'nnight,
Time's pace is so hard that it seems the length of
seven year.

ORLANDO
Who ambles Time withal?

ROSALIND
With a priest that lacks Latin and a rich man that
hath not the gout, for the one sleeps easily because
he cannot study, and the other lives merrily because
he feels no pain, the one lacking the burden of lean
and wasteful learning, the other knowing no burden
of heavy tedious penury; these Time ambles withal.

ORLANDO
Who doth he gallop withal?

ROSALIND
With a thief to the gallows, for though he go as
softly as foot can fall, he thinks himself too soon there.

ORLANDO
Who stays it still withal?

ROSALIND
With lawyers in the vacation, for they sleep between term and term and then they perceive not how Time moves.
As You Like It. III. ii.

Men cannot control time, but they can strive to live symbiotically or flow with the river of their particular time. Their states of mind and activities, whether flowing with, battling against, or navigating through time, influence how they perceive or fail to perceive and understand the possibilities and challenges of the time and place with which they must contend. Here as elsewhere in As You Like, the purest of Shakespeare's great comedies, time always reverts to comic time.

Time remains comic because the play's characters eventually find ways or have the good luck to seize the opportunities their moments afford. Rosalind understands that timing, grasping the moment, is the key to success in love, as in all things in life. Timing, she maintains, is also a proper measure of a lover's trueness of heart:

Orlando1

My fair Rosalind, I come within an hour of my promise

Rosalind

Break an hour's promise in love? He that will divide a minute into a thousand parts and break but a part of the thousand part of a minute in the affairs of love, it may be said of him that Cupid hath clapped him o' th' shoulder, but I'll warrant him heart-whole. IV. i.

For lovers in Shakespeare, as for his comic heroes, the present moment is all. The present is the place where joy and happiness live, the place where decisions can be made, the place where the future can be determined. That makes the present the only place where we can freely choose our fates and create who we are and will become:

> And therefore take the present time
> With a hey, and a ho, and a hey nonino, V.ii.

As You Like It 's citations on time are typically Shakespearean, they are both profound and self-evident, as is the case with the following often memorized passage from *Julius Caesar*. Though here the modes switch to tragic and the qualities of time darken to make universal Rosalind's insight that grasping the propitious present moment is key to success in all things while inability to act dooms one to failure:

> *There is a tide in the affairs of men.*
> *Which, taken at the flood, leads on to fortune;*
> *Omitted, all the voyage of their life*
> *Is bound in shallows and in miseries.*
> *On such a full sea are we now afloat,*
> *And we must take the current when it serves,*
> *Or lose our ventures.*
> Julius Caesar— IV, iii

Opportunity, romantic, political, professional, or economic, is the rare gift of time. But Time, like a mercurial lover, must be grasped in her chosen moment or lost forever. As Andrew Marvel, rising to a Shakespearean moment, put it:

> *But at my back I always hear*
> *Time's wingèd chariot hurrying near;*

And yonder all before us lie
Deserts of vast eternity. To His Coy Mistress

The terrible finality and irreversibility of time rears its head in life's critical junctures in each of Shakespeare's romances as well as his comedies, histories, and tragedies. Like Lear's beloved Cordelia, once a splendid opportunity is gone, it comes no more, never, never, never, never, never.

Brutus' observation ends with a dire warning that those who ignore an opportunity for good fortune will be chastised with bad fortune. The chastiser is that strumpet Fortune who always turns with a vengeance against those who spurn her unrepeatable gifts. Wise Ulysses echoes this theme in one of Shakespeare's most cynical plays, *Troilus and Cressida*:

For time is like a fashionable host
That slightly shakes his parting guest by the hand,
And with his arms outstretch'd, as he would fly,
Grasps in the comer: welcome ever smiles,
And farewell goes out sighing. O, let not virtue seek
Remuneration for the thing it was;
For beauty, wit,
High birth, vigour of bone, desert in service,
Love, friendship, charity, are subjects all
To envious and calumniating time. **I.*iii.***

Here time is a devouring monster that eats up all our past good deeds so do not expect favor for noble deeds past. Time is like a fickle host who goes through friends as a prodigal goes through gold. Time is an envious, slanderer who has no respect for friendship, past service, beauty, wit or virtue.

Shakespeare gives fickle, pagan time and Fortune backhanded Biblical reinforcement from Jesus's parable of the good steward, which exerted an enormous influence on Shakespeare, Elizabethan Protestantism, and English shopkeepers. The moral of the parable is that we must make good use of our opportunities, time, and resources or lose favor with God. The prize goes to those with the mettle and initiative to assume the risks of failure:

Who choses me must give and hazard all he hath. Merchant of Venice II, vii

Or, as Hamlet put it with pithy profundity, "*The readiness is all.*" V, ii

The theme of the worthlessness of unused talents and fortune's payback for ignored opportunities appears again in a much more densely worded passage in one of Shakespeare's most complex plays:

Thyself and thy belongings
Are not thine own so proper as to waste
Thyself upon thy virtues, they on thee.
Heaven doth with us as we with torches do,
Not light them for themselves; for if our virtues
Did not go forth of us, 'twere all alike
As if we had them not. Spirits are not finely touch'd
But to fine issues, nor Nature never lends
The smallest scruple of her excellence
But, like a thrifty goddess, she determines

Herself the glory of a creditor,
Both thanks and use. *Measure for Measure I.i.*

The title of the play, like this speech, signals the moral of the parable of the talents and the stewards: that we are to be judged and rewarded on the basis of how well we use the measure given us. Unused talents, ensconced in a closed heart, the drama soon shows us, will be replaced with correspondent vices at the first real temptation. Likewise, unused opportunities bring ill fortune in their wake. And so at the end, its wise and thrifty Duke Vincentio, who represents the playwright and the deity, turns all the play's nuptial mishaps into opportunities to realize marital bliss.

Just as the tide of men's affairs can rise to a flood, it can ebb leaving us stranded on a bank and shoal of time.

The time is out of joint—O cursèd spite,
That ever I was born to set it right! Hamlet I, v

Hamlet's rotten state of Denmark is an example of time as a distinct moment or time field that is a realm unto itself with its own character, challenges, rules, and atmosphere. *Hamlet's* dilemma and task define the Hamlet tardis. The tardis itself is Hamlet's tragedy which lies not in flawed character but in a dilemma for which there can be no happy resolution.

An earlier character, and progenitor of Hamlet as well as Macbeth, who is seriously out of sync with time is Richard II:

Music do I hear?
Ha, ha! Keep time. How sour sweet music is
When time is broke and no proportion kept!
So it is in the music of men's lives. Richard II V, iii 41

Men's lives are like musical compositions; we compose them as we live. They can be grand, harmonious and beautiful like a symphony, or cacophonous when the time is broken and no proportion kept. Richard II's own decisions, his deficient judgment and defective sense of proportion led to blunders that have broken time for him, though he lacks the self-knowledge to accept responsibility for his mistakes. Unable to embrace his tragic end with dignity, he evades it with self-pity. For Hamlet the chief blame for breaking time lies in a situation not his making, in his uncle Claudius killing Hamlet's father, the king, to marry his mother, the queen. But other leading tragic figures in Shakespeare, Lear, Macbeth, Othello, Brutus and Richard II, have themselves broken the time and thereby turned the music of their lives into cacophony.

Time in Shakespeare is alive with multiple possibilities that unfold and grow from mere seeds. 'There are many events in *the womb of time* which will be delivered,' Iago promises his gull Rodrigo. But try as we may, not even Iago, can reliably predict which seeds of time, will flower, wither, fail to grow at all, or be devoured by some hungry fowl. "If you can look into the seeds of time and say which grain will grow and which will not," says Banquo to Macbeth. The seeds of time are the potentialities of the present moment; only time itself will reveal which seed will sprout and what may grow from it. The best we can hope for, as

Hamlet realized, is to be ready to avoid snares and grasp opportunities when they appear.

Antony and Cleopatra shows time as decay fed by satiety leading to lost opportunity and ending in disaster and death. But all is not lost, for the two heroes, especially Cleopatra, are redeemed by the power of their art and artifice, as evoked in the poetry of Shakespeare. Full of glorious hubris at the outset, Antony declares:

> Let Rome in Tiber melt, and the wide arch
> Of the rang'd empire fall! Here is my space,
> Kingdoms are clay: our dungy earth alike
> Feeds beast as man; the nobleness of life
> Is to do thus: when such a mutual pair,
> And such a twain can do it, in which I bind,
> On pain of punishment, in the world to weet
> We stand up peerless. V. i.

These grand lovers have chosen to live for romantic love, absolute and reckless, and become, thereby, two of the great romantic icons of all time. Antony places love above power and sacrifices power to love in repeated impulsive decisions and heedless actions. Oblivious of the future, he and Cleopatra become the Macbeths polar opposites living in their present, idolizing their past. In the end Antony waveringly accepts his unsought sacrifice as he affirms the transcendent value of his love for himself as well as for Cleopatra. In her own more elegant and more conscious, though equally self-dramatizing, way Cleopatra follows Antony's affirmation.

Give me my robe, put on my crown; I have
Immortal longings in me. Now no more
The juice of Egypt's grape shall moist this lip.
Yare, yare, good Iris; quick. Methinks I hear
Antony call. V. ii.

Throughout Shakespeare's opus love creates value that transcends power and wealth, affirms life, and vindicates honor. Love is time's only worthy opponent. *Antony and Cleopatra* is a celebration of the value of love, life, and art over the protagonists' glorious folly, and against dull Octavius's relentless connivance and cold lust for power.

Love as an archetype of transformation is ever linked with the yearning for sacrifice and metamorphosis in romance and in life, the ultimate sacrifice being annihilation or death as in *Romeo and Juliet*. However, in *Antony and Cleopatra* Shakespeare attempts through his heroine a symbolic transcendence of death and the fatal irreversibility of time proclaimed at the end of *King Lear*. "Symbolic" needs to be stressed because for Antony and Cleopatra transcendence is only a fantasy. It is as if Shakespeare gave them a chance to save face in what appears to be an instance of his falling in love with his characters, a habitual and wonderful sympathy that enabled him to create so many glorious, iconic dramatis personae. The credibility of Cleopatra's challenge to time the destroyer will depend on each viewer, and that is probably as Shakespeare wanted.

With *King Lear* and *Antony and Cleopatra* juxtaposed Shakespeare gives us two seemingly opposed, yet somehow complementary, perspectives which serve to deepen our felt understandings of death, tragedy, and time. Antony's claim to transformation is, however, less credible than Cleopatra's who always maintains her dignity, and in the end is transformed by that dignity. Antony's actions at the end and throughout exhibit greater instability, often bordering on self-caricature, and sometimes sinking to buffoonery. He has lost more than Cleopatra. He has lost manhood while she becomes more thoroughly the irresistible female siren, as such she even wins the praise of Enobarbus who began as her harshest critic.

Unlike in *King Lear* and Shakespeare's other major tragedies, the viewpoints of the key characters in *Antony and Cleopatra,* except Octavius, appear to waver from the play's beginning to its end. Just as it enhances their humanity by making them multidimensional, the wavering gives us and them varied and often conflicting yet interdependent perspectives. Ultimately, it creates a *metastance* to our initial perceptions and judgement. Metastance is where mature conceptions of truth and life usually abide in Shakespeare's later plays. Old Yorick was a man of infinite jest, Hamlet tells us. Their creator was a man of infinite compassion, in addition to infinite jest, infinite variety, and seemingly infinite wisdom. His compassion for his characters generated the *metastances* that allows each their unique truths.

Though the experienced playwright who wrote *Antony and Cleopatra* is much older than the passionate young poet of the Sonnets, he remains infused with his youthful hope that true art can transcend time. *Antony and Cleopatra* embodies a vision that art can defeat death by capturing immortal beauty or, as Jung might say, by embodying an archetype in compelling and accessible form. Enobarbus comes to realize that Cleopatra captures the beauty of a mature woman which is no longer physical but lies in her all conquering charm, wit, and guile. She has made herself a work of art:

> Maecenas. *Now Antony must leave her utterly.*
> Enobarbus. *Never; he will not:*
> *Age cannot wither her, nor custom stale*
> *Her infinite variety; other women cloy*
> *The appetites they feed, but she makes hungry,*
> *Where most she satisfies, For vilest things*
> *Become themselves in her, that the holy priests*
> *Bless her, when she is riggish.* II.ii.

Antony is not bound to Cleopatra by sexual attraction so much as by the appetites of the imagination, mind, and soul which only increase with feeding. She can transform the ordinary into ornament and even holy priests bless her divine artistry. Through her artistry, Cleopatra can make death itself, time and fortune's ultimate weapon, the instrument and occasion of her triumph over time by achieving her own apotheosis:

> My desolation does begin to make
> A better life; 'tis paltry to be Caesar:
> Not being Fortune, he's but fortune's knave.
> A minister of her will; and it is great
> To do that thing that ends all other deeds,
> Which shackles accidents and bolts up change; V.ii

It is often remarked that *Antony and Cleopatra* is full of character change and imagery of change. In the denouement, Cleopatra finds the remedy that will shackle accidents and bolt up change to secure her immutable fame. Yielding to immortal longings, Cleopatra gains immortality as an iconic lover and woman of infinitely varied charm, wit, and guile. Shakespeare leaves it to her and Antony's great adversary Augustus Caesar to proclaim as a matter of fact their timeless fame:

> She shall be buried by her Antony,
> No grave on earth shall clip in it
> A pair so famous V. ii.

Beginning with *Twelfth Night* and into the later plays, often called the romances, time the destroyer becomes time the problem solver, the healer, and the facilitator of reconciliation. Viola, *Twelfth Nigh*t's ambisexual heroine, relies on Time to extricate her from her dilemma:

> O Time, thou must untangle this not I;
> It is too hard a knot for me t' untie. II.ii

The play's wisest and best characters, Feste and Viola, live in harmony with time and must help Orsino, Olivia, and Sebastian, who are out of tune with the time, through times' untangling into a final harmony represented by matrimony. Malvolio, the iconic comic butt, is a perfect example of someone out of harmony and always at odds with the time. Like Macbeth, Malvolio is obsessed with future ambitions and rejects his present station.

Where conflict is primarily external, as in the early comedies, place more than time shapes the world of a Shakespearean play. Time seems hardly present in *A Midsummer Night's Dream* where it loses its way amidst a haze of fairy dust. The often morally dark places of hot, sunny Venice in *The Merchant of Venice*, set the play's initial tone. That tone changes radically in Belmont, a morally improved, moon lit world where the moment dissolves into a romantic eternity to the music of Jessica and Lorenzo's "In such a night." Time creeps into Arden forest, a magical place that dispels time's power to harm. *Twelfth Night*'s Illyrian time creates another magical place where fragmented identity can be healed and restored. Illyrian time anticipates time the regenerative power of the last plays where, as a healer, it attains equal prominence with place the refuge.

In the *Winter's Ta*le Time moves beyond its personification in Fortune to become an actual character, representing the playwright himself. It's as if the writer were to come on stage to discuss and explain his play, because it fails to fully explain itself. An innovative dramatic strategy, and probably necessary due the limits that audience comfort impose on the length of a play. The action occurs in two different timeframes separated by sixteen years and two generations, and it transpires in two separate countries. Thus, Shakespeare decides to beg the audience's pardon for his odd artifice:

> Now take upon me in the name of Time,
> To use my wings. Impute it not a crime
> To me or my swift passage that I slide
> O'er sixteen years, and leave the growth untried
> Of that wide gap, since it is in my pow'r
> To o'erthrow law, and in one self-born hour
> To plant and o'erwhelm custom. V.i.

What is most significant here is that Shakespeare, who in the Sonnets saw Time as the great enemy and destroyer, reconciles himself to his old adversary and prepares to embrace and dramatize its power as a healer. Time's healing is achieved by overwhelming custom and law to allow the movement from a diseased, hate blighted time into a new era. With time's renewal, a new generation, by letting love bloom in itself, sparks love's resurrection in the older generation and its triumph in the play.

Forgiveness initiates the resurrection of love in Shakespeare. *The Winter's Tale*, a comedic contrapuntal to the springtime world of *Romeo and Juliet* where bad timing brings romantic tragedy, stages an improbable and illusory reversal of time's tragic irreversibility. In the only physical resurrection in his opus the statute of Hermione comes life reversing all the sorrow her apparent death has caused:

> Music, awake her: strike! [Music.]
> 'Tis time; descend; be stone no more; approach;
> Strike all that look upon with marvel. Come,
> I'll fill your grave up: stir, nay, come away,
> Bequeath to death your numbness, for from him
> Dear life redeems you. You perceive she stirs:
>
> [Hermione comes down]. V.iii,

Lear tests for Cordelia's breath with a looking glass then a feather, which if it stirs will redeem all the sorrows he ever felt. He dies searching her lifeless face for signs of the resurrection that never comes. But Hermione, long thought dead by her husband, King Leontes, was never dead. Time's reversal, Shakespeare's thematic juxtaposition tells us, is an illusion possible only with stagecraft artifice. *King Lear*, for Shakespeare, has the final, unchallengeable say on death, time irreversible, and the nature of the gods.

Nevertheless, in The *Winter's Tale* time creates the miracle of forgiveness and redemption for the older and the younger generation. That miracle becomes the theme of Shakespeare's final play, *The Tempest*. Time the destroyer becomes time the redeemer, who redeems through art and thereby creates something new and strange—love and wisdom where there was selfishness, cruelty, blindness, and folly.

Full fathom five thy father lies;
Of his bones are coral made;
Those are pearls that were his eyes:
Nothing of him doth fade
But doth suffer a sea change
Into something new and strange.

The Tempest. I. ii.

Time in *The Tempest* is inscrutable, magical, miraculous, and redemptive. Time itself becomes the universal artist. Indeed, time becomes God the Artist as Redeemer. For Shakespeare true redemption from what is base, dull, and ugly in life comes through art. Prospero's island is a once desolate place transformed by his art which fills it with music and a beauty that even ugly, defective creatures, like Caliban, appreciate:

The Isle is full of noises,
Sounds and sweet airs, that give delight and hurt

Not,
Sometimes a thousand twanging instruments
Will hum about mine ears, and sometime voices
That if I then had wak'd after long sleep,
Will make me sleep again;

The Tempest III. iii

To a more accomplished extent than with any other Shakespearean character, an extent preeminent in all of literature, Prospero embodies Jung's wise old person archetype. Prospero is what anthropologists refer to as a mana personality which emanates a supernatural ability or charisma with magical healing powers. The wise man, or wise woman, has achieved individuation and manifests mature wholeness. Activation of this archetype launches the transformative process of initiation through which the young initiates, Prospero's daughter Miranda and her swain Ferdinand, establish their readiness for the next stages of life's journey. Shakespeare's final statement on Time from Prospero reveals the bard's own identification with the wise man archetype, and affirms the value of life's evanescence. It is a prime candidate for the most sublime passage ever written.

Our revels now are ended. These our actors
As I foretold you, were all spirits and
And are melted into air, into thin air:
And like the baseless fabric of this vision
The cloud-capp'd towers, the gorgous palaces,
The solemn temples, the great globe itself.
Yea, all which it inherit, shall dissolve,
And, like this insubstantial pageant faded,
Leave not a rack behind. We are such stuff

As dreams are made of, and our little life
Is rounded with a sleep. IV. i

Shakespeare's majestic poetry defies summary and beggars all analysis, but that hasn't stopped the critics. Here is my attempt: "the pageant's disappearance provides Prospero with a simile to illumine the transience of man's life, the fragility of identity, and the universality of dissolution. As he moves from the dissolution of the masque to that of the actors in the play of life who will melt into sleep, Prospero teaches that art, the world, and identity are as ephemeral as the stuff of dreams. In The Tempest man realizes transcendence in dissolution and subsequent transformation into a higher consciousness or metastance. When Ferdinand masters Prospero's instruction, he can accommodate dissolution and death, and thereby gain spiritual transformation and rebirth. Through rebirth a man becomes fully alive to his essential nothingness, and the conflicting stances that his ego identity assumes are superseded by the transcendent unity of metastance." Metastance is Shakespeare's corollary to Jung's coincidentia oppositorum, In Shakespeare you can always hear the resonance of opposites. The Shakespearean resonance of opposites or metastance, is the bard's response to the harsh dualities that fill the existential challenges of time.

And so Shakespeare's works dramatize the dualities of time, its glories and terrors, opportunities and reversals, triumphs and disasters. They give us warnings and wise advice by poignant, incisive examples. All in all, they offer us a uniquely, comprehensive vision and a catechism for understanding the God who rules all, Time Itself.

MEDITATION VII: DUALITIES IN THE SPIRAL OF TIME

1. Life and death is the most momentous duality of our existence. Sustaining life is our supreme challenge, and death is the greatest mystery. Time, as a conscious individual perceives it, unfolds from the womb and fades or ends abruptly with death, while varying in its apparent speed and qualities throughout life. Our lives are instruments that measure time. All instruments and their measures are unique. Time's passage intensifies in moments of fear and anxiety, slows in rest, speeds when we rush, and appears to stop in some sleep states, mystical experiences, or momentary ecstasy. Deep sleep, mysticism, and ecstasy appear to take us out of time—but we always awaken into time. We think death takes us out of time. The living may speculate, yet none have experienced Hamlet's "undiscovered country from whose bourn no traveler returns." We cannot help but wonder and fantasize about that mysterious country, have hopes for and fears of it, see ghosts and other apparent visitors from it. Yet our fate does not allow us to verify our speculations to the satisfaction of science. For the living it remains the great unknown and unknowable.

2. An afterlife for a soul that persists once the body perishes would entail a degree of complexity, a layering of realities in the universe, or at least a duality of matter and psyche-spirit, that science presently lacks the

necessary sophistication of theory and reach of experimentation to grasp or verify. Such concepts as the psyche, spirit, and soul may or may not ultimately prove useful to a more advanced state of science. Their validity may rest on whether we regard the universe as monistic, simply physical or material, or see it as dual, psychical as well as physical. Jung and Whitehead both opted for an organizational dualism of matter and psyche, but rejected ontological dualism with its Manichean moral implications.

By traditional philosophical understanding, time is dual, all is past or future: in the moment there is nothing. Just as light from far galaxies reveals only their past billions of years ago, the light, sound, and sensations that our brains receive from our immediate environment are also from the past, although a much more recent past. So time, as we perceive it consciously, is either a reflection of or reaction to things past or an anticipation of things to come. Some parapsychological claims suggest a more complex view of time with pluralistic rather than binary dimensions. However, the dual nature of time, the time built into the structures of our western languages, finds parallels in much of what science tells us about the world.

Science shows us chirality, or right and left handed molecules, positive and negatively charged particles within atoms. We know life tends to bifurcate with two sexes and with chiral, imperfectly symmetrical creatures with two, four, six or eight legs and two eyes and two ears. The earth and magnets have two poles, bodies subsist on a hot and cold scale, there is light and darkness, up and down, subject and object, past and future, phenomenon and noumenon, time and space, energy and matter.

Scientists speculate about matter and anti-matter. Our computers systems are based on binary operations. The fundamental unit of human society, the mother and father, is binary. No wonder that we think and construct languages in terms of dualities. Consciousness as a dimension can be paired with the physical universe, mind with matter. Consciousness mirrors, complements, measures, and enhances our knowledge of the physical universe, and it directs many of our interactions with it. These facts argue by analogy that psychic and physical are a basic duality of our universe.

The bifurcate universe does not allow us to establish the soul's immortality by grounding it and mind in eternity, because the universe's psychical and physical aspects, like all its dualities, remain knowable only in time. We know empirically that the universe has duality running through it, but the idea that the temporal universe itself might be paired with a separate eternity is pure speculation. If the universe of time needs to be paired with eternity, do not time and eternity need something to pair with, and so on into an infinite regress? Moreover, the soul cannot be changeless, eternally outside time, or it would cease to have meaning within time.

The Hindus and Plato may have sensed the need to keep the soul within time in order to keep it meaningful, which gave them an additional reason, beyond cultural tradition, to embrace metempsychosis. The doctrine of Eternal Recurrence, may stem from the same concerns. Another approach to immortality is Plato's ancient concept of archetypes, patterns that persist across and throughout time. Jung embraced the archetypes, but took no position on the immortality of the soul despite pressure from religious people. Likewise, Whitehead said his philosophy is "entirely neutral on the question of immortality." The immortality of the soul is a metaphysical concept. It is not the same thing, Jung and Whitehead each realized, as an afterlife for which there appears to be interesting, though inconclusive, empirical support. Certainly, the universe that science now apprehends is vaster and more complex than what Plato, Archimedes, or Aristotle could know. We can only speculate, imagine, dream, and hope that our ideas about the human condition in the great beyond will not ultimately prove empty speculation, but instead hold something of great constancy that can endure and indeed prevail amidst the swirling vagaries of the great spiral of time.

3. It is often remarked that time is linear and moves like an arrow, a common sense notion accepted largely, though not entirely, by classic as well as modern quantum physics. Others, such as Vico, Nietzsche, James Joyce, and physicists who speculate that the universe may consist of cycles of big bangs and big busts, have stressed the circularity of time. Psychologically, their emphasis is understandable given that life on earth runs in cycles, birth thru death, the seasons, day and night etc., and the circle is a basic pattern of nature just like the line. (Joyce plumbed the existential impact of time's irreversibility more profoundly in his early story *The Dead* than in his more theoretical explorations on time in *Finnegan's Wake*.)

There is another basic pattern whose analogy to time has been less often explored, the spiral. (One may conjecture that the circle symbolizes stasis, the line flux, and the spiral a process.) Galaxies display a spiral movement as do hurricanes, cyclones, and other movements down to the drains in our bathtubs. The double helix itself resembles a spiral.

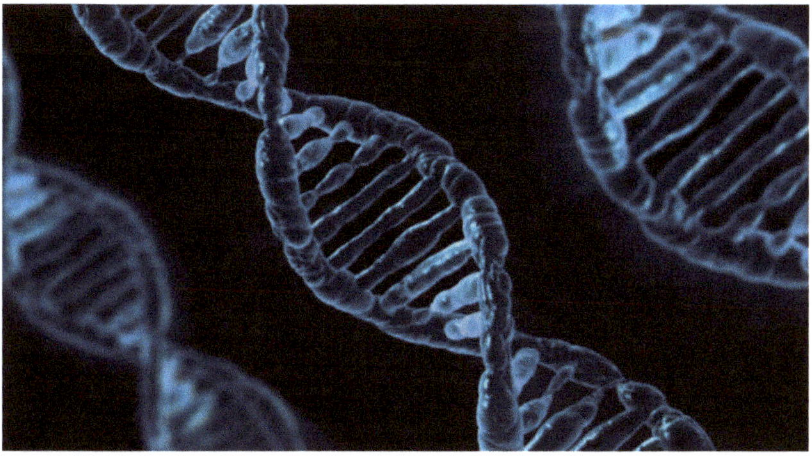

If we ever isolate the strings physicists theorize about, we may find they form spirals. Unlike the line and the circle which are singular in form, the spiral, like time itself is dual. It becomes dual by combining the line and the circle. A commonplace truism is that history repeats itself. Some wags say that history does not repeat, it rimes. That also suggests a truth: history repeats but always with significant variations. The spiral circles, but it is actually moving toward someplace new, like the arrow, though not on a straight, predictable trajectory. Each cycle varies, each part of the spiral is different, its movement wavers and vacillates. These variations are creative and filled with novelty and, for Whitehead, adventure.

Pattern repetition altered and enriched by novelty, which the spiral can symbolize, is a fundamental characteristic of time and the unfolding, creative universe. Whitehead develops this concept in *Process and Reality* which, of course, presents process as the ultimate nature of reality.

Whitehead's own thought has been compared to a spiral, revisiting key notions from different perspectives. The spiral serves as a metaphor for the creative advance of time, process, and divinity in Whitehead, as well as in this work. Jung and others also have noted that the unfolding spiral characterizes and/or symbolizes God working in the process of time. In his poem *The World* the seventeenth century metaphysical poet Henry Vaughn conveys the magnificence of the great, galactic spiral of Time:

I saw Eternity the other night,
Like a great ring of pure and endless light,
All calm, as it was bright;
And round beneath it, Time in hours, days, years,
Driv'n by the spheres
Like a vast shadow mov'd; in which the world
And all her train were hurl'd.

Vaughn's image of eternity and the temporal process presents a vast celestial icon for The God in Time. Had he the benefits of modern astronomy, Vaughn might have recognized that The God in Time has stamped Its spiral image and symbol billions of times across the entire created universe. When advanced conscious beings gaze out into space and eternity, they can see everywhere the spiraling symbols of The God in Time. Looking into the microcosms of their own little worlds and themselves, there too they will find The God is Time everywhere, driving, limiting, and shaping all.

4. Every individual thing is part of, interconnected with, some larger whole. The Nile is part of Africa, Africa is part of the earth, the earth is part of the solar system, the solar system part of the galaxy, there are

clusters and superclusters of galaxies and so on. Each moment is part of the whole of time, but is time as we know and experience it part of a larger super-temporal reality? How does time relate to eternity, does eternity transcend time, is eternity real or, as I have suggested, just an intellectual construct like the zero? Eternity is often viewed as time's opposite. But if time is the ultimate reality, eternity must be the whole of time, not its opposite. Of course, no one can know the whole of time because no one knows the future. Truth, value, creativity, and meaning are only possible where there is time. If we could get out of time, that would put us in an anarchic realm where there is no truth or meaning or future. To want to go there is nihilism.

5. It's a commonplace that time is dual and all things in time are bifurcated, since if there is nothing in the moment, all is either past or future. Or to put it differently, each present moment is defined by its past and in respect to its future. The over-riding principle is the temporal process itself that incorporates the duality of past and future into the present moment which contains nothing in itself, but is the dynamic nexus linking everything before with everything after. That means the temporal process, contains both the infinite one of "eternity," or time in its entirety, and the infinite multiplicity of the past and future conjoined at the uniplural nexus of the present. Each present moment, as the makers of the *I Ching* understood, has its own unique character, meaning, and value.

Nature may abhor a vacuum, but it appears to love dualities. Is reality itself bifurcated between mind and matter or conscious and unconscious realms as it is between what was and what will be?

Philosophers and psychologists speculate about the relationships between consciousness, or mind, and matter or body. Some always try to reduce one to the other, yet none achieve a final resolution satisfactory to all, and the bifurcation remains a fundamental aspect of our existence like past and future.

6. *For every action there is an equal and opposite reaction*---Contrasting Isaac Newton's third law with Nicolas Carnot's second law of thermodynamics is thought provoking. Many physicists believe that the second law means that the state of entropy of the entire universe, as an isolated system, must always increase over time. This would entail a natural tendency in any isolated system to degenerate into a more disordered state. Might Newton's third law indicate that there may be an opposite and, over time, equal counterforce to entropy? Or, does the dualism we see throughout the universe indicate that destructive entropy must have a creative counterforce towards increased order? Does the destructive force lodged in matter have a creative counterforce lodged in mind or psyche? Just as things fall apart because of entropy, some things, particularly organic entities, have a tendency to coalesce into systems of increased order. One function of mind is to discover and impose order to counter the natural entropy of matter. Newton, an intense believer in the Christian deity, might have supported this train of thought, as did Whitehead who said, "It lies in the nature of things that the many enter into complex unity." Physicist Paul Dirac interpreted his famous equation to mean that for every particle there exists an antiparticle, exactly matching the particle but with opposite charge. Does this apply to basic principles so that entropy must be complemented by an opposing principle

of coalescence? Or to use Whitehead's similar term, *concrescence,* which he defines as a growing together of a many into a unity of a one. Teilhard de Chardin refers to the same general phenomenon as *complexification.* Concrescence or coalescence or complexification would be, then, the universal engine of creativity that continuously brings novelty to birth to counter entropy and balance destruction.

As entropy increases and the universe cools, creating areas of moderate temperature, conditions become more favorable to organic entities. Under these conditions the drive toward greater order, or coalescence into new unities, will likely accelerate and become more consequential. The principle of coalescence appears to apply sporadically and unpredictably in species evolution. Similarly, while they age, that is fall subject to physical entropy, humans through individuation have opportunities to expand consciousness or coalesce toward a greater order that creates novel wholeness—e.g. Prospero and the wisdom of age. Evident in life, coalescence may also offer a general explanation for psychic phenomena such as synchronicities and telepathy which can mark as well as advance the individuation process. Coalescence might eventually avert entropy's heat death of the universe. It has billions of years to find ways to do so.

Time is fundamentally dual so that time the destroyer, that is entropy, should be matched by time the creator of novelty time the agent of coalescence. Coalescence has religious implications. It can evidence the creative activity of Time as God to counter and eventually overcome the tendency of entropy to push the universe toward ultimate anarchy. It, thus, portends to turn St. Paul on his head as the rise of man displaces the fall of man as the prime narrative of humanity. Ancient dualisms held that two irreducible contending forces accounted for all that exists, their beliefs may be seen as mythic representations of entropy and coalescence.

7. The dualistic features of time and the universe spawn a host of difficult, albeit interrelated, questions for the philosophically minded.

*Are meaning and value byproducts of something more fundamental about the universe, such as a God in Time who created them, or the principle of coalescence, or concrescence, or complexification?

*Did time first appear when consciousness dawned somewhere in the vast tumult of matter and energy? In other words, does time, like meaning and values require a consciousness to perceive and interpret it?

*Was the universe dual, physical and psychical, from its beginning or did one state arise from a predecessor state to create duality, as God in Genesis created Eve from Adam's rib?

*Modern physicists speculate that in the first moments of creation matter, energy, and the extent of the universe were quite different than they are today. Was time also different then? Did the universe change time, or does time slowly change the universe?

*Does time reflect more primal elements and laws that gave the universe birth and propel it forward? If so, what is more primal than time? Whitehead speculated that extension may be more primal since it includes time and space.

* Or, is time itself the most fundamental force, the dynamism of the universe creating, shaping, and enlivening it with meaning and value,

a force that contains past and future but is so basic that it has no counterpart to itself?

*Are time and the universe one inseparable, interactive, uniplural reality? Are time and the universe the hieros gamos of cosmic deity?

*Is God time, or is Time god, or both?

You will find my answers to some of these questions herein, others are more heuristic, springboards for those who wish to explore further.

MEDITATION VIII:
MULTIPLE UNIVERSES

Is time like the DNA of the Universe or is it the life of the universe or both? If the universe, or a sub-universe, loses time or is without time, does it die or is it dead? The answer is that a universe without time would be a dead universe because it would be changeless, static, and without a glimmer of life or consciousness. In short it would not be a universe at all, but rather a vast nothingness. Time is the life of the universe, there could be no real universe without time, the precondition and medium of being. Without time everything would fall into a figurative black hole where all differentiation is lost. Without time there is no consciousness, and without consciousness there is no logic. Any universe that can be thought about must be subject to both logic and time. It might not be the case, however, that every universe that can be thought about would have our periodic table.

Just as different species and creatures have their own unique DNA, but all have life, so time fields, or tardises, within time have their unique patterns, or DNA if you will. Different sub-universes, or different eras or areas in our universe, might have different temporal patterns, what are called laws, yet still be ruled by Time. Possibly, there is a meta-universe, that includes ours as a sub-universe, consisting of many sub-universes all sustained and enlivened by time, each subject to logic, and each with its own metaphoric DNA. Indeed, the discrete, yet interacting, time fields of our universe may sometimes seem/act like sub universes.

In the modern era scientists and philosophers began to speculate more frequently about multiple universes, though such thought experiments have been common since Spinoza, Descartes, and Leibnitz. The concept of multiple universes reflects our perceptions of time fields, time junctures, and time frameworks. It has been said, that each new moment creates a new universe. Physicists speculate that the universe a minute after the big bang was essentially a different universe than what existed at the big bang's initial flashpoint, or what exists today. They further speculate that the laws of the universe, began to change and take shape in the earliest timeframes of its inception. The laws were not fixed at the very beginning, but were created and defined as the universe unfolded in time. Conscious agents are similar to the universe itself in the sense that we create a new timeframe with each decision we make, a view that accords with Hannah Arendt's assertion that each person has the potential to change the world.

Is time the process that unifies or makes the universe a universe? Time as process can elude the ancient dilemmas created by the flux versus stasis debate of Heraclitus, Zeno and others. However, process is more difficult to grasp philosophically than is stasis or flux. Any who doubt should try reading A.N. Whitehead's *Process and Reality*. The chief paradigms for process are from biology and human life, and like life itself they are often nuanced. Time as process is existential, it is in all places and things with which we must contend. The simplistic paradigms for stasis and flux appeal to logic. But as Hamlet reminded Horatio, "There are more things in heaven and earth, Horatio, Than are dreamt of in your philosophy." The one thing that is everywhere in heaven and earth is time.

What if our universe consists of many, perhaps an infinite number, of sub universes? The sub universes would be subsumed in the greater universe because each would have a common characteristic, time, and each would be subject to logic. Time might appear to run in odd stops and starts, but still be time. In Stanley Kubrick's "Barry Lyndon" seventeenth century time runs slow. Its pace is definitely and deliberately different from time in "Clockwork Orange" or the time on a spaceship in "2001." Kubrick's three movies portray three different time tardises. Dreamtime, whether in Aboriginal myths or in the dreams of Jung's patients, presents another set of tardises where the time order and pace are shaped by active archetypes whose effects that can't be measured by watches, clocks, and calendars.

Much has been said here of time, but little of space. Whitehead, reflecting Einstein's space-time, thought that both are subsumed under extension. Mine is a different perspective. Time is the active force and space the stage where it acts. Time is dynamic, it is the great destroyer and creator. Time acts in and through space, which makes space passive. The expansion of the universe across time, creates space. Because time is the dynamo of existence, I identify time with God, the supreme creative force. God is concerned with existential time, the time conscious beings experience. Existential time does not change; it is the same for Nefertiti,

Cleopatra, Eleanor of Aquitaine, and Melania Trump. Reliance on physics for our understanding of time has a major disadvantage: physics is an ongoing endeavor changing with time itself. Who can say that we have reached or neared a final physics? A millennium more of scientific endeavor may well lead to a physics more radically different from today's than today's is from Newton's.

Beauty seems timeless: Nefertiti, Queen of Egypt, 1353-1336BC

The role of time in the universe is critically important, so let me repeat, as in a spiraling summary, saying in slightly different ways what has been said before: Without time there could be no consciousness to perceive the universe or to distinguish between things within it. Time is a necessary condition for conscious thought, since without time nothing would be separate from anything else and consciousness depends on perceiving relationships and making distinctions between separate things and events.

These considerations lead to a critical conclusion: ***The universe is everything subject to time and time is its ultimate reality. The God in***

Time is the God of the universe. What if there were two or more separate, totally unrelated universes subject to time? To know that would require a single consciousness cognizant of both universes. That consciousness would create an over universe with a supertime where the separate universes would become sub universes each with their own subordinate time, like tardises or time fields. There may be sub universes, but this universe is the only one we have and can know, and its supreme law is time—or super time. Time unifies the universe and thereby makes it a universe.

Some philosophical, religious, and even scientific thinkers maintain that God is outside the universe imposing order on it and interjecting value into it. The scientists argue that because matter and anti-matter are equal and symmetrical in the universe they should annihilate each other. They don't: therefore, something outside of matter and anti-matter, a God, must be preventing that. However, if God is interacting with the universe to sustain it across time, that would put the deity in the universe. God need not be outside the universe to sustain its order, order can be/is an aspect of the process that is God and time. The speculation that deity is outside is groundless and useless. When asserted by religious thinkers seeking to place the deity in eternity, it becomes a rationalization to save Him from being tainted by the flaws of the universe He created. They fail to realize that He does not need to be saved by their rationalizations. Attempting to do so is a kind a blasphemy that places them above God in judgment of Him.

MEDITATION IX: TARDISES AND SYNCHRONICITIES

1. An enthusiastic student of the I Ching, Jung focused on both the quality and quantity of time, whereas scientific time in the West focuses chiefly on the quantity of time. I will use the term ***tardis*** to designate the qualities of a specific temporal context forming a discrete continuum of time, or a timeframe or field or span of time that has its own unique and intrinsic qualities and character. A tardis is like a magnetic or gravitational field, only operating within time, whose organizing forces are activated archetypes. A tardis would contrast with an arbitrary span of time: for example, the Civil War might be seen as a tardis while 1850 to 1853 would only be a span of time. The term's meaning should not to be confused with the name for time machines in the science fiction series *Dr. Who*, whence the name is borrowed. A related term might be the German *umwelt* or the world as experienced by a particular individual or organism. We might speak of the *umwelt* of the Civil War, but would not likely speak of the *unwelt* of the Presidency of Millard Fillmore. Another relevant German term is *zeitgeist* which refers to the spirit of a span of time or tardis.

Unlike an *umwelt* but like a *zeitgeist*, a tardis is independent of the individual agent. It is, however, interdependent, as a distinct section or field, with the larger reality of the web of time. Nonetheless, each tardis is unique because it has its own dominant complex of archetypes, limits, and a beginning and end even though they may be shadowy boundaries that phase into adjoining tardises just as gravitational fields phase into each other. Tardises are temporal analogues to Whitehead's concept of spatial "societies" or distinct orderings of spatial perceptions. Tardises are distinct ordering of temporal perceptions. Whitehead's term "societies" refers to interrelated, interdependent groups whose members have a common character that they mutually reinforce, as is the case with a tardis. Tardises or societies of events contrast with aggregates where the particulars lack significant interconnections with each other, such as a pile of debris. Events in tardises form a distinct temporal ecosystem or process.

Entering into a tardis resembles an episode in time travel. Seen spatially a tardis might look like an island: e.g. Sicily is part of Italy and Europe but distinct from them in important respects. The temporal context of a tardis might be seen to resemble a highly irregular honeycomb where the tardis is a single cell. Another way of viewing a tardis is like a bubble in time, which may appear larger (or smaller or at least different) inside than it would appear to be to an observer outside the bubble—or in a different tardis. There are sub tardises or bubbles within larger tardises, as well overlapping tardises, and tardises of varying strengths and lengths of endurance, just as there are gravitational fields of varying sizes and strengths overlap within the Solar system or any area of space. Time as change appears to move at different speeds and in divergent patterns within the tardises. Whitehead's term "pulsations" might be used to characterize the distinct speeds and patterns that distinguish tardises. Indeed, tardises are analogous to dynamic pulsations in the wider temporal universe.

The temporal universe of tardises may be likened to a vast bubble bath, the Hamman of God and/or the gods. Figuratively, each discreet era may be seen as a kind of tardis. Bubbles burst, and so may tardises burst in time as bubbles do in space. The tardises of new world civilizations appeared to burst upon contact with the West. The tardises of China and India merged slowly, and with somewhat less catastrophic results, with that of the West. The tardis of the Islamic world, whose history with the West is fraught with conflict, has begun to violently resist merger with the West.

These historical patterns give us clues about our possible responses to extraterrestrial contact, an event that would inaugurate a major new tardis. It could be a catastrophe, as presented in sensational science fiction conquests, a sudden transformation to something better, or a slow but bumpy evolution, or a long war of attrition as is playing out between the West and Islam, or a peaceable interpenetration and/or a slow, unrecognized absorption. The one thing it seems least likely to be, is the noisy, violent, physical combat dominated by a lone action hero of superhuman physical prowess and sexual attractiveness, as portrayed in Hollywood space invasion movies.

The tardises are not merely conceptual constructs any more than Ireland is merely a conceptual construct. They have their own character and protocols as well as peculiar tones, qualities, and ambiences all part of an interrelated web of great intricacy and complexity. You may not know exactly when you phase in or phase out of one, just as you can't always be sure if you've left one region and entered its neighbor or you can't be sure when you've left one sphere of influence to enter another. Like geographic regions and spheres of influence the character of a tardis is usually best seen and understood from the areas near the centers. The centers, where the dominant archetypal forces are strongest, appear to exert a pull like a gravitational or magnetic field, only in time rather than in space.

In terms of archetypal psychology, tardises occur when highly activated archetypes coalesce or configure to create a time field. Powerful archetypes can warp human existential time according to Jung, even as massive objects in space warp space-time according to Einstein. Seen another way, the coalescing of archetypes might be a psychic version of what happens with organic chemicals. The archetypes act like the atomic forces of various elements which bond chemically to form organic compounds of extraordinary complexity. A major difference being that these archetypal compounds in the tardises are more fluid and unpredictable. The tardises are ever changing in response to alterations in the activation levels of their archetypes and other archetypes in the temporal vicinity. Jung referred to the phenomenon as archetypal configuring. However, the formation of temporal tardises involves binding connections which would be defined more accurately as a coalescing. The archetypal coalescing takes place in the psychic rather than the physical realm. While it is something of greater constancy than an incidental gathering, it is more difficult to define than an organic chemical bond.

History appears to contains time fields, tardises, pulsations that work similarly to gravitational and magnetic fields or spheres of influence. These time fields shape the ways historians organize their material. In history the influence of cardinal figures, e.g. Caesar, Jesus, Mohammad, Napoleon, Darwin, who constellate archetypes, create veritable fields, or tardises, in the events around them. The same holds for major wars, disasters, inventions, movements etc. The cardinal figures parallel Jung's personified archetypes, while wars, movements etc. resemble archetypes of transformation. But are the time fields real, in respect to the outer physical and psychic realms, or are they and their histories merely conceptual devices the mind employs for perceiving and ordering these realms? Does this question make any difference? Or is what matters the way the tardises operate in our lives, not what they are? The way they operate is like an ecosystem where events, patterns, and trends interact with, limit, enhance, and shape each other.

Identifiable time fields or tardises, possess independent, discernible characteristics, and are very roughly analogous to organic chemical compounds or gravitational and magnetic fields. Analyzing them and the extent of their sway might facilitate predictions about their future and understanding of their past. Jung speculated that powerfully constellated archetypes can create dynamic fields that effect the physical world through psychic and synchronistic phenomena. At a mundane level, understanding the constellated archetypes can help guide psychotherapy, as dreams and synchronicities can, and light the path to individuation. Questions, speculations, and of course skepticism, abound in this area.

2. *Unus Mundu*s refers to the context, the inherent interconnectedness of all being, and to the field or medium in which the archetypes act. Time itself is the *Unus Mundus* sought by the alchemists and Jung. *Unus Mundus* as a concept resembles Hindu ideas of Brahman and is sometimes symbolized by the Anthropos or cosmic man in alchemical and esoteric writings. It finds contemporary expression in the idea of Gaia, though the realm of Gaia is only the earth, not the universe. Known through intuition more than reason or sensation, *unus mundus* and its cognates provide a metaphoric expression and framework for acausal concepts of meaning such as synchronicity, teleology, implicate order, and *unio mystica*. Jung called them acausal, but these concepts might instead refer to forms of causality different from the linear, physical causality of mechanistic science. Or they might refer to an inherent parallelism between the psyche and the physical world. The pivotal idea behind *unus mundus* is the interconnectedness and interdependence of everything. For Jung healing, or restoring wholeness to the psyche, involves establishing a connection with the *unus mundus*. Wholeness brings knowledge of God, whom we encounter first through the self and Who ultimately unfolds in the *unus mundus*. Healing puts us more in sync with time itself, and at peace with The God in Time.

3. Jung's concept of synchronicity posits an acausal connecting principle that relates events to reveal meaningful connections bridging the psychic and physical worlds. *Synchronicity may be defined as: a*

simultaneity conjoined or fraught with psychological meaning that discloses significant qualities of a moment or period of time. Synchronicities are occasions that manifest the interaction and/or parallelism of the subjective with the objective. We notice a synchronicity when we see a meaningful coincidence between external events and states in a perceiver's psyche. For example, if someone dreams of a tiger and the next day encounters an actual tiger in an unexpected place. Phenomena that the scientific mind assumes linked only by chance or coincidence, may be meaningfully linked, Jung held, by synchronicity.

Primitive and ancient peoples were fascinated by synchronicities. They tried to interpret them through traditional divination by omens, portents, and augury. Jung sought scientific explanations for synchronistic phenomena. The result was a difficult, though intriguing, concept that became one of his more controversial ideas. Jung's theories on synchronicity are, I believe, best viewed as attempts to achieve a deeper, more comprehensive understanding of 1) the complexity of time and causation, 2) the interrelationships of mind and matter, and 3) the unique qualities and structure of moments in time, 4) how constellated archetypes of the *unus mundus* can cause events and shape understanding of them. Synchronistic thinking reaches across the structures of mechanistic time and causation to show us that constellated archetypes can give us unusual experiences that shed meaningful light on significant problems in our lives. These meanings yield novel insights that are gifts of the self to the conscious ego for the purpose of opening opportunities to advance individuation. When we notice synchronicities, it is likely because an active archetype pre-consciously organizes our train of thought.

Rational, mechanistic approaches to time and linear causality do not exhaust the meanings of temporal interconnections. They focus on one conception of time, the arrow of time, that scientifically trained minds have learned to deal with exceptionally well, and therefore tend to rely on. Modern man's globe transforming success with science, industry, and technology seems irrefutable proof of our prowess in rational thinking. However, science and technology require specific cultural and philosophic contexts and constructs that simply did not exist before the Greeks, and did not emerge as factors until the Renaissance-Reformation era. Earlier cultures turned inward to more spiritual or magical approaches to the major challenges of people's lives. Rather than investigate a problem scientifically and lay out their options, they might do a divination or consult an oracle or make a sacrifice to a deity in search of meaningful coincidences that could illumine their situations. The *I Ching*, which is still widely used, is one of the best known and most credible of these systems of augury. Augury and science have the same underlying objective, prediction.

We cannot dismiss as invalid these early approaches relying on unconscious guidance simply because so far human beings seem less successful using them than using the scientific-rational approaches. Other higher intelligences in more advanced worlds than ours may not depend on science and technology to the degree that contemporary human cultures do. They may have moved beyond or even abandoned this way, or possibly assigned it to A-I. We know that the human mind and our genetic platforms have developed a multiplicity of ways of surviving and thriving in the world, science being only one and until recent centuries not a very pervasive or consequential one. Quantum physics tells us that our universe is more complex and multifaceted than can be easily grasped by the mechanistic thinking characteristic of traditional Newtonian approaches to physical reality. Indeed, we have no more idea of how complex the universe may ultimately be, or how varied the forms of intelligence within it are, than we know how the most advanced beings in our Milky Way's sister galaxy Andromeda spend their vacations. Synchronicities, along with psychic phenomena, may at the current point in history seem mysterious, but it is a premature to dismiss them as improbable chance events lacking significant meaning for their observers.

4. Life hands us lessons; with synchronicities the self draws attention to lessons the ego wants to ignore. Jung believed that the synchronicities in his life and those of his patients offered information that could help solve problems, advance individuation, and facilitate healing. He saw them as messages from the deep self that divulge significant patterns or facts that otherwise the rationalistic mind would reject or dismiss.

Jung saw a metaphysical implication in synchronicity, that is an interconnectedness of events in the psyche and in the physical world. It exhibits the universe's essential duality of macrocosm and microcosm, physical and psychical, which is for Jung its uniplurality manifest in the *unus mundus*. Uniplurality and duality are complementary, not mutually exclusive, terms. They refer to multi-centered fields, duality being bipolar and uniplurality being multipolar. The essential interconnectedness of the uniplural *unus mundus* entails a basic interdependence, or binding, within the universe itself which Jung, Whitehead, Bergson, Teilhard de Chardin, the mystics, and others have seen as a fundamental religious insight, and some believe may be the single most important religious insight.

Another suggestive approach is to see synchronicities as instances of the spontaneous coalescence that counters the basic entropy of the material world. In this way they would remind overly rationalistic moderns that there are more things in the universe and this world than are recognized by our mechanistic and positivistic philosophies.

A skeptic might object that nothing, no force counter to material entropy, causes the synchronistic events. The mind merely locates events whose correlation to concerns of the psyche enables the psyche to use them to try to send a message to an oblivious ego the way dreams do. It is clear that certain people, those in Jungian therapy for instance, are more likely to notice synchronicities and scrutinize their dreams. An individual psyche so energized may notice many synchronicities, just as a psyche attuned to certain types of beauty or threats will notice more of them. Only a few dreams are highly significant, most are routine rehashes and comments upon the events of the day. Likewise, only a few synchronicities are highly significant; but we ignore these, Jung believed, at our expense and peril.

Jung's explanation for significant synchronicities, that highly activated archetypes constellate corresponding events outside the psyche, is close to making mind a counter entropy. Activated archetypes, Jung believed, deepen our perception of time through synchronicities, thereby exposing the very qualities of periods in time and revealing events as far more complex and interrelated than what our rational scientific thinking assumes. Synchronicities show the psyche and the *unus mundus* to be more interconnected, and interdependent than rationalism can fathom. The double duality of time, the past and the future along with its absolute irreversibility and interconnectedness, are at once the ultimate irrationality and the ultimate reality with which we must cope. While time may seem irrational, we sense the *unus mundus* of time and of physical and psychical experiences within and across time. Synchronicities, as well as other psi events, serve as reminders that there are things beyond rational understanding that, nonetheless, are real, significant, and valuable. Our sense of the *unus mundus* does not dispel the inexplicable, but it allows us to see synchronistic events not as threats but as potential guides and helpers.

MEDITATION X: TEMPORAL CONCEPTS

1.WISE PLANNING. Shakespeare's Prospero embodies the wiseperson archetype and is the epitome of wise planning, just as King Lear is the very embodiment of foolish, irresponsible planning. Jesus's parable of the stewards illustrates the need for and critical value of wise planning. From a Jungian perspective, wise planning is the coordinated application of intelligence, feeling, and intuition to manage time, a strategy required for dealing with life in the *unus mundus*. It involves recognition of the limits time imposes on consciousness, like birth, growth, maturation, decline, death and of course time's irreversibility. Fear of death and the wish to deny the powers of time or the refusal to adjust to change and decline fosters psychological resistance to planning. Those who refuse to plan assume that, by living for the moment and in the moment, they can somehow evade time and death. They are impelled by an ego driven hubris, a quest for self-sufficiency which is tantamount to an unattainable absolute independence. They strive to evade time and escape or deny reality to death. They refuse to reconcile with the self's essential interconnectedness and interdependence with the *unus mundus*. Wise planning entails the humility to recognize and accommodate mortality and interdependence. It allows us to employ imagination to turn life and time into art. Those who refuse to plan consciously let the conflicting forces of the unconscious and the passions of inflated ego do their planning for them. To counter ego inflation, the psyche may intervene by activating archetypes that compensate the ego's refusal to plan for the inevitable or to correct ego's bad planning. Archetypal interventions come in the form of dreams, startling events, unusual psychic phenomena, and synchronicities.

2. TIMEFRAMES. Humans perceive events within the contexts of *timeframes*, a generic term that includes but is not identical with tardises. Tardises refer to temporal structures with specific qualities, force fields around constellated archetypes, or distinct ecologies, within the continuum of time that are independent of the particular observer. Whereas timeframes are the observer's way of consciously organizing time. Like tardises, timeframes often phase into each other and overlap. A memory intruding from a past timeframe can color, shape, or correct how we perceive events in a current timeframe. Similarly, conscious and unconscious anticipations of future timeframes can influence our perception of the current timeframes. Memories of past timeframes along with anticipations of future timeframes shape and can even determine decisions we make in the present.

Time itself is a continuum without pause or apparent beginning or end, in time we are always *in media res*. However, our minds seek to organize and structure events to find meaningful patterns (to impose a teleology) that can guide our decisions, shape our hopes and apprehensions for the future, and structure our past to make it a reliable and useful guide to the present and future. The mind works like an artist painting a picture, a writer telling a story, or a musician composing a song or symphony. His frame is an essential structuring device: consider the *Mona Lisa*: our perception of it would be altered and our appreciation of it might be lessened without a frame. Where the painting begins and ends would seem ambiguous, its meaning would be harder to fathom. The artist's impulse to frame his painting is nearly universal. We may not judge a book by its cover, but we do judge a novel, play, or movie, by its time frame, its beginning and end, just as our judgement of a picture is shaped by its frame.. Frames create a limiting context for pictures, they tell us that the picture captures a moment. By setting limits, the frame helps define the artist's vision of the structure, meaning, and value of that moment within the frame.

In life the contexts that timeframes establish are critical to shaping individual values, and decisions. Conscious beings live within shifting timeframes set by their desires, fears, hopes, and cultural expectations. A mother may interpret events in the timeframe set by her concerns, and aspirations for her children. The priority of being a mother can shape the timeframes she constructs in order to give meaning to her life and help her make decisions. The artist, the businessman, and the scientific researcher, all have professional priorities that shape their timeframes. Most of us plan for a career, a family, a relationship, college, vacations, or retirement and in many cases death. These plans establish interacting timeframes that we rely on to guide decisions that give direction and meaning to our lives.

2. In writing a history, we customarily set our chosen subjects within the timeframes of their eras. The medieval, Roman, or Renaissance eras serve as timeframes that shape and color our views of events within them. Historians construct their histories within an era or age much as artists construct a painting within a frame or writers frame their narratives. Timeframes are critical to military, political, and business planning. In the 1930s the French built the Maginot line relying on a timeframe that extended back through World War I, not forward to account for advances like the tank and new strategies such as blitzkrieg. Germany, relying on a more contemporary timeframe, seized the opportunity afforded by French atavism to simply run its attacking armies around the impregnable French line.

Political and business strategies and decisions also fail because they rely on timeframes oriented too much to the past or the present and too little toward the future. A political or business leader can enjoy great success because his timeframe is realistic and future oriented: one thinks of leaders like the US founding fathers, or business titans like Andrew Carnegie, Henry Ford, Bill Gates, or Steve Jobs. Montgomery Wards' Sewell Avery brought disaster on that company because, waiting for a depression that never came, he missed opportunities to update and expand. A wise political, business, or career decision depends heavily on identifying an accurate and appropriate timeframe as a context for that decision, a timeframe cognizant of the past yet rooted in the present while looking to the future.

Arnold Toynbee believed civilizations fall into decay and collapse because they rely on timeframes locked into the past that prioritize preserving their identity. Enamored of their own past, they lose their ability to adapt to new challenges. Businesses also fail, like civilizations, because their timeframes are oriented toward crisis management and the short term bottom line. For example, CEOs neglecting to plan beyond the next quarter or politicians whose time horizons end at the next election pursue short sighted policies that will in the long run damage or destroy their companies or countries. Short term political timeframes have encouraged accumulation of vast debt burdens in contemporary democratic societies, as well as the degradation of our natural, infrastructural, and cultural environments. Short term timeframes encourage unsustainable policies that end in disaster because they postpone dealing with urgent longer range problems until it is too late: as Hamlet said, "the readiness is all." The wastrel, as a stock character in novels and dramas, exemplifies the disastrous consequences of myopic time frames. But societies too become wastrels. When they fall, vast numbers of hapless citizens fall with them.

Are time cohorts fellow travelers in timeframes or denizens of specific tardises? Are junctures in time accidental, or do they reflect tectonic shifts in the underlying archetypes that shape tardises. Do time beacons and peak experiences reflect archetypal forces that, like magnetic poles, set up and order events in their timeframes or tardises? Can we predict the future or understand the past better through studying the nature and dynamics of tardises and timeframes? Do our destinies make us, or do we make out destinies? Nothing can generate big questions like time.

3. TIME COHORTS. We move through our lives grouped in overlapping timeframes that often exhibit the qualities of tardises. Our contemporaries are time peers who share key life stages. Born in close time proximity, we pass together thru childhood, adolescence, youth, maturity, middle age, old age, very old age, and death, thru school, marriage, careers, child rearing, sicknesses, and retirement. We share the same big events in the world, the nation, and locally at the same stages of life. Today 9/11 would seem quite different in the eyes of a 10 year old than in 30, 50, 70, or 90 year old eyes. We feel a special kinship with our time cohort, they are in effect our time family or village. The further apart in time we are from others, the fewer events we share until we reach a point where no discrete timeframes overlap. Then people are no longer time neighbors, but denizens of alien temporal countries which grow more distant as they go further back in time. The worlds of Joyce, Dickens, Shakespeare, Dante, Virgil, and Homer seem to grow progressively more alien and shadowy even as their denizens human qualities can remain amazingly vital and immediate to us. So it is that time separates each from all, and yet through underlying duration amidst change can unite one with all. It is the nature of the temporal process to accommodate both change and duration, two contradictory forces that are essential to each other, and whose contention shapes our lives.

4. JUNCTURES IN TIME. Tardises, and timeframes as well, are often separated by junctures, these can range from sharp breaks like fault lines to subtle shifts like inflection points. They include routine events everyone experiences such as birth, death, marriage, job, and residence changes. The unusual or even extraordinary events that do not occur

regularly in most lives, such as winning a Nobel Prize or catastrophic accidents or diseases, make some of us stand out from the crowd of our time cohorts. In the context of history, the lives and achievements of seminal leaders like Buddha, Christ, Mohammed, or Caesar, Elizabeth I, and Lincoln can mark key junctures. The crucial junctures can also be wars, disasters, discoveries like the printing press, the steam engine or the atom bomb, or social movements like mass migrations, industrialization, or the 20th century entry of women into hitherto male ruled workplaces and the associated economic, political, and sexual revolutions. Such major changes can initiate a new tardis where ways of thinking, feeling, and perceiving are altered significantly among large and/or key segments of the population.

The junctures can inaugurate or terminate eras, or if you will tardises. The life of Mohammed activated archetypes that began and sustained an era, as did the life of Christ. Often eras phase in slowly with a shift in the direction of a curve, like the industrial revolution, rather than emerge abruptly such as happened with the French Revolution, the Civil War, and World War I. Time junctures and the eras or tardises that follow them can be named after events, e.g. postwar Europe, or even a person without whom a break in time would not have occurred, for example the Elizabethan era in Britain. Crucial junctures can initiate new tardises wherein the previous patterns and sea currents of events change dramatically. When that happens they can transform our intellectual framework for perceiving what we regard as reality. They can also initiate entire paradigm shifts in science, the arts, technology, religion, or politics.

The Darnley portrait of Elizabeth I (1575) shows her remarkable intelligence and cool command

5. TIME BEACONS AND LANDMARKS. Our perception of time is curved by momentous events that become inflection points spawning new tardises in the continuum of time, just as space is curved by gravity fields of massive objects. The closer in time or in space the greater the pull momentous events exert. For example, time in American history was curved by the Constitutional Convention, Ft. Sumter, Pearl Harbor, 9/11 attack, and some hope and others fear by the election of Donald J. Trump. Major events can stand like beacons or quasars to the temporal

worlds, or tardises, that surround them. A death, an earthquake, a hurricane, a war, a financial crisis, a plague etc., may or may not mark divisions between eras, tardises, or timeframes, but it can still form a landmark in the course of time.

Along with timeframes and junctures, time beacons and landmarks are patterns or structures in the flow of time that the mind uses to navigate through and impose meaning or order on events. Kant referred to the organization within consciousness of individual events, or the mind's organizing of raw perception, as the transcendental unity of apperception. Its function includes imposing, identifying, creating, and discovering meaning within time to achieve, control, or influence, and to understand, or at least explain, past events and try to predict and plan for our futures. We seek guideposts that resist the flux of time. Time beacons are guideposts that help us navigate time. They can give us a fleeting sense of control, if never mastery, over the raw flux of tempestuous and unruly time.

In a religious view, especially a providential Christian view, the time beacons often become signs of divine intervention in human affairs. The Christian concept of *Kairos* refers to an ultimate time beacon, the redemption of mankind marked by the appearance on earth of Christ, the Son of God. However, other major events, such as the Reformation, are likewise time beacons for Christians, many of whom also see the founding of the state of Israel as a crucial time beacon. Muslims have a separate set of time beacons, and rather different attitudes toward the founding of

Israel. Modern physics has proposed its own ultimate time beacon, the big bang which marked the creation of the universe and time itself.

There arose a widespread feeling in the West in the early 20th century that we are approaching a once in a millennium, supernova time beacon. In the most quoted, and possibly the most emblematic, poem of our era, the prophetic voice of W. B. Yeats gives an iconic statement of this hope, cloaked as it often is in dire apprehension.

THE SECOND COMING

Turning and turning in the widening gyre
The falcon cannot hear the falconer;
Things fall apart; the centre cannot hold;
Mere anarchy is loosed upon the world,
The blood-dimmed tide is loosed, and everywhere
The ceremony of innocence is drowned;
The best lack all conviction, while the worst
Are full of passionate intensity.

Surely some revelation is at hand;
Surely the Second Coming is at hand.
The Second Coming! Hardly are those words out
When a vast image out of Spiritus Mundi
Troubles my sight: a waste of desert sand;
A shape with lion body and the head of a man,
A gaze blank and pitiless as the sun,
Is moving its slow thighs, while all about it
Wind shadows of the indignant desert birds.

The darkness drops again but now I know
That twenty centuries of stony sleep
Were vexed to nightmare by a rocking cradle,
And what rough beast, its hour come round at last,
Slouches towards Bethlehem to be born?

What is the modern era's defining second coming? The candidates are many. Our era has seen the rise of ominous and formidable political forces like totalitarianism, globalism, environmentalism, and mass genocide. We face the challenges radical Islam poses to the West and all rival religions. We have experienced life changing breakthroughs like antibiotics, jet travel, the internet, smart phones, and the atomic bomb. Massive social changes confront us, like the liberation of women from traditional economic constraints, the sexual revolutions, the extension of human longevity and the ensuing population expansion, a communications revolution creating a massive planetary nerve system, rampant global urbanization, and the resulting widespread environmental degradation.

Or do the most consequential changes lie unnoticed within religion itself, like the spiritual revolution that started in Bethlehem two millennia ago? Are they in the decline of the old religions and their loss of credibility in the face of startling scientific discoveries, and in their dire unmet needs for adaptation or even reinvention? Do they lie in the limitations and outright failings of traditional morality in modern life, or in the rise of ideological fanaticism that strives to displace religion? Will we see a spiritual counter reformation to the scientific revolution?

Perhaps the signal change is too small to be recognized at this point, as was Christianity in its first decades. Is it already out there in a mustard seed of thought in a yet unwritten work by an unknown author somewhere in Africa or East Asia or US flyover country? Is the center that cannot hold the unbalanced, orthodox Trinitarian Godhead or the vaunted omnipotence of the Islamic ego-god or both? Is the rough beast slouching toward Bethlehem a new revelation of Godhead? The forces that bind great civilizations, like galaxies of culture, are losing their hold while entire arms spiral off into the unknown. A widening gyre becomes a spiral: Yeats the poet-seer sensed the spiraling nature of time and human history. Is Yeats widening gyre an inflection point for the great spiral of the human adventure itself whose next major phase may be somewhere off the Earth in space?

6. INTENSE OR PEAK EXPERIENCES become time beacons in the lives of individuals. Core values and underlying attitudes, or changes therein, can arise from the heightened awareness created in these life defining events. The road to Damascus, the burning bush, sighting Christ on the cross or performing a miracle are Biblical examples. Indeed, such experiences play critical roles in conversions to Christianity. These events are seen as revelations or signs from God, divine interventions if you will, or they can work like an artistic epiphany, to use Joyce's term. The ultimate Christian manifestation of such an experience is Kairos, God becoming man in Christ. It forms the defining event of the religion.

But outside of religious experiences extraordinary events can shape values along with philosophical and political positions and personal priorities. Hamlet's confrontation with his father's ghost and Macbeth's encounters with the weird sisters are notable examples in literature of the impact of extraordinary events on personal decisions and priorities. The authenticity of a philosophical, political, or religious stance can be judged and evaluated by the depth, intensity, universality, and authenticity of the experience or set of experiences upon which it is based. An overwhelming experience becomes iconic for those who make it the foundation of their faith no less than for the person who actually had the experience. Beliefs are always a leap of faith, but not all leaps of faith are equally authentic. Beliefs based on weak, routine, unreflecting, or fraudulent experiences, or on just going along with the crowd, are inevitably less compelling and less persuasive. Certainly, they carry lessened moral force.

Episodes of psychosis can also become powerful life defining, peak experiences. It is sometimes an open question whether peak experiences are borderline or even outright psychotic episodes. Skeptics certainly will raise this possibility with such events in the lives of Mohammad and some of the Christian saints. Psychotic episodes can be authentic, and authentically destructive, though sometimes destruction is necessary to clear a path for new good. This applies not just to episodes of individuals. The most dangerous manifestations are when whole societies are ruled by sociopathic crowd psychology instigated by ideological demagoguery like Nazism or Stalinism. On the other hand, when revolutions clear away what is diseased or useless, their long term benefits can outweigh the damage from their excesses. Peak experiences are not good or bad, but as we make them so. Evil and destructive events can have their authenticity as well as good.

Intensity and complexity of experience are important measures of art in all its forms. *King Lear* is often regarded the supreme tragedy or even the greatest work of literature because of its profundity, rich complexity and unique intensity: that is, the intensity of the experience of Lear himself and of the play's most perceptive, sensitive, and thoughtful viewers and readers. Intensity of experience shows depth of character and authenticity in literary constructs such as Lear, Hamlet, Macbeth and Othello. Complex characters, like these, display multiple levels and qualities of intensity which contribute to their depth, authenticity, and memorability. We evaluate the character and authenticity of people in our lives in terms of their breadth and quality of experience, their defining intense experiences, and their ability to grow and learn from, and communicate these experiences.

One can, of course, develop depth of character and exceptional judgment from very limited but highly authentic experience that is deeply felt and seriously reflected upon, as Anne Frank or the author Jane Austen illustrate. Moreover, breadth or value of experience does not guarantee great or even good character, as our politicians so often demonstrate. Learning requires more than just experience. The sight of Christ on the cross or the horrors of the holocaust mean everything to some and little or nothing to others. Some cannot learn in the best of circumstances from the greatest of experiences, others learn from the worst and the least. Ultimately, we must rely on our own best experiences if we are to achieve authenticity, and gain historicity, by making the time that is given us our own.

7. DESTINY AND FREE WILL. Destiny is a force that sets a general course for the future that resists change by the actions or will of individuals. Destiny is determined by a power within time, the will of the gods, historical forces, or an irresistible archetype in a time field or tardis. An ineluctable destiny, or fate, implies that individual free will cannot alter

the course of events, and it may even imply that human free will does not exist. Some maintain, to the contrary, that Heisenberg's uncertainty principle establishes the possibility of free will and limits destiny to probabilities. Heisenberg's principle has caused many to ask the old questions: is time fixed and static? is it a flux? or is it in fact a creative process?

Fate, or to use the Greek term *moira*, excludes randomness and nullifies human free will. In cultures where fate reigns supreme, e.g., where every problem is explained away with "God wills," the *creative continua* of the culture abate alongs with progress. While destiny can be good, bad, or indifferent, fate has negative connotations of impending disaster, a sentence of doom, or irretrievable loss. The concept of all powerful fate entails a static universe and engenders fatalism which becomes an excuse for inaction leading to spiritual torpor. A static conception of time raises destiny from a probability to a certainty which, historically speaking, is usually identified with the irresistible will of God or the gods.

By contrast, nothing can be destined in a flux universe other than the flux itself. Flux denies destiny by subjecting all to randomness. Total randomness would appear to make freedom meaningless by denying continuous identity to individuals, although Sartre and other flux oriented existential philosophers resist this conclusion. For them radical freedom, far from being meaningless, becomes the most significant meaning available to man. In a process universe a bad destiny can be altered by either random events, or conscious action. The idea of a process universe with free will and a God unfolding in the *creative continua* limits destiny to a tendency or a built-in probability which can be altered by other forces. Chief among these is the will to change the course of time with free decisions that create authentic individual identity, build character, and thereby shape one's own destiny. A Jungian term for the will to shape destiny and create one's identity is individuation. Process philosophy also gives a decisive role to the individual free will. The unpredictability built into a process universe is a critical factor for freedom: the process of time becomes an interplay between apparent randomness and tendencies with various probabilities. The ultimate meaning lies in neither randomness nor probabilities, but in the process, including the action of free will, that mediates between them.

Heidegger, in a provocative insight, construed destiny and freedom as symbiotic. He defined them in respect to an individual's existential time or historicity, as distinct from an absolute time set by either the omnipotent, omniscient, perfect God of Christian orthodoxy or the mechanistic Newtonian time (or Einsteinian space-time) of science. For Heidegger, time's basic tense is the future: reaching out from the future time turns back to assimilate the past which has created the present. One's destiny, one's personal timeframe is finite, it has a beginning and an end in death, and there is no immortal soul of course. The authentic individual must accept as his own the history or past that has been given him, and he must choose to be responsible for it. That means we are to enact freely in our time the destinies that are given us by no choice of our own. Therein, Albert Camus' absurd hero is born. Not everyone, perhaps only the strong, can live in the mode of destiny which requires facing and accepting our ineluctable nothingness in death. For Heidegger, Camus, Hannah Arendt, and other existentialists owning one's destiny takes strength of character, but it makes freedom possible. Their position harkens back to the ancient Stoics who held that while we cannot control our fate we are free to choose our attitude toward that fate. We can own our fate instead of fate owning us.

Some speculative physicists take a position exactly opposite to Heidegger's, Camus's and Arendt's, that in our universe everything in time is fixed and our movements through time, our distinctions between past and future, are illusory. Their viewpoint amounts to radical stasis, but from a scientific rather than a traditional religious starting point. While theirs is just a theory, if it were to become a widely accepted doctrine the social and psychological consequences could be devastating. If all were determined from beginning to end, none could justly hold anyone accountable for past actions or even current decisions. Hitler and Stalin would be as guiltless as Anne Frank. Moreover, none could take pride in their achievements or feel shame or guilt for their misdeeds. We could have no hope or ambition to improve ourselves or the world. If time is an illusion, so is progress, free will, and responsibility; without belief in them civilizations flounder.

Martin Heidegger in his study

Baruch Spinoza, the great rationalist-monist philosopher and civil libertarian hero is also at the opposite end of the philosophical teeter totter from the existentialists on destiny, free will, and time. Spinoza's logical rigor is seductive, but it's a temptation that must be resisted because empirical reality resists and rebels, just like the existentialists, and most important like life and the self. While Jung shares a pantheistic outlook with Spinoza, his concepts of individuation and wholeness resemble Heidegger's ideas on historicity, freedom, and destiny. Jung might see Spinoza as a paramount example of reliance on a single function of the psyche, thinking or logos, at the expense of ignoring the others, especially feeling and intuition, or at least subordinating them to thinking.

A monistic fixation on thinking impedes individuation for Jung, as it would impede authenticity for Heidegger. Spinoza does illustrate, perhaps better than any other great philosopher, the tendency of rationalism to disregard the creative function of the dialectic between thinking and its opposite feeling, as well as ignore intuition in order to facilitate embrace of pure logos and monistic stasis. Spinoza sees the identity of nature and God without recognizing that both are a process. For Jung thinking must coordinate with the other functions to create an interdependence that furthers the individuation and thereby can realize true freedom. By over identifying with any one psychic function, the ego corrupts that function and isolates it from the self. Isolation from the self impedes growth toward the deeper wisdom, manifest in the wise person archetype. It stifles the soul and with it life itself. For Jung true freedom comes from listening to and accepting guidance from the self: man's destiny is to realize freedom through the self.

MEDITATION XI: PROCESS, FLUX, STASIS

Look to the past and the universe seems a stasis, because the past cannot change. Look to the future and it seems an oceanic flux full of vast, unknown possibilities for change. Moreover, the past is there to remind us the future often did not happen the way we expected. Look at the present and it seems the knife's edge of a mysterious process wherein an unknown, fluid future ineluctably changes into a fixed, known past. Is reality more like a flux or a stasis, or is it a process? These are three different ways of viewing the universe and time. Time is all three, but some may argue that reality and the universe itself exist only in the present and so must be a process. Others would contend that only the past and/or the future are real, because the present is merely a moving boundary separating them.

The three perspectives have generated a range of different philosophic positions. Process philosophy tries to integrate all three perspectives, toward the aim of wholeness or integrity, yet process stresses the present which we can enjoy. "Enjoyment" is a critical term for Whitehead who sees it as God's fundamental aim for sentient beings. He defines enjoyment by intensity, harmony, and complexity, wherein it has an oblique relationship to beauty. Stasis philosophers, in their quest for rational certainty, focus on the past, which we can know. Flux philosophers focus on the future, which we can change, because they seek to motivate political change. These are not hard and fast categories but sections of a spectrum that ranges from extreme stasis to extreme flux, with most philosophers finding some congenial niche in the temperate zones between the poles. The following Table presents my speculations on where influential thinkers fall in respect to the basic positions on time:

STASIS	PROCESS	FLUX
Parmenides, Plato	Bruno, Boehme	Heraclitus, Protagoras, Democritus
Plotinus, Origen	Lactantius, Occam	Pelagius, Montaigne
Augustine, Aquinas, Kant	Hegel, Schiller, Goethe	Hume, Mill, Russell, Wittgenstein
Spinoza, Leibnitz, Hobbes	Whitehead, James, Pierce, Bergson, Dewey, Heidegger, Hanna Arendt	Marx, Sartre, Foucault, Derrida
Newton	Einstein, Darwin	Hawking
Calvin, Barth, Buddha?	Jesus, Tillich, Teilhard	Kierkegaard
Mohammad	Cardinal Newman, Zoroaster?	Mani
Thales, Aristotle?	Jung, Erich Neumann	Freud, Hillman, Skinner
Dante, Swift, Tolstoy, Eliot, C.S. Lewis	Shakespeare, Goethe, Milton, Mann, Joyce	Dostoevsky, Melville, Beckett, Brecht, Kafka

William James, a seminal process philosopher, associated stasis with a tender minded personal psychology, and flux with tough mindedness. Process, lying in between, is more practical, commonsensical, and realistic, at least in the eyes of process philosophers, who in the last century and a half tend to be Americans. Temperament, climate, and geography can influence philosophical positions no less than culture, personal experience, and traditional assumptions about time.

Stasis philosophy tends to view the universe as fundamentally monistic, rational, eternal, and perfect. Spinoza may be the purest of the major stasis philosophers, but Plato is the Godfather of stasis. Stasis is congruous with the Son as Logos in the Trinity, the ego, the thinking function, absolutism, monotheism, monarchy, and authority. Christian orthodoxy derives its doctrines of *summum bonum* and *privatio bon*i and its comedic final outlook from stasis philosophy. However, orthodoxy's one-sided philosophical commitment to stasis forces its opposite, flux, into the shadow where it challenges stasis with an implicit dualism that posits Satan as the adversary of God and a powerful agent of anarchy. Manichaeism and other radical dualisms manifest this dynamic.

A flux perspective, in contrast to orthodoxy's stasis, will be pluralistic, favor polytheism, dualism, or atheism, and view the universe as ultimately irrational and imperfect. Flux is the stand of the rebel and the anarchist, just as stasis is the stand of the conformist and the reactionary. Flux philosophy draws on and bolsters the Satan-Lucifer-Prometheus (Prometheus being the positive face of destruction) and shadow archetypesl along with reductive thinking, relativism, ideology, and what Erich Neumann called the ethic of rebellion. Flux's characteristic outlook tends to be irony, whereas stasis has a comedic outlook, and process tends to be epic, but includes a tragic sense of life.

Stasis holds evil to be ultimately unreal and treats it as an impurity to be blamed on scapegoats like Lucifer, or Prometheus and driven into the wilderness. Flux is inclined to see evil and irreparable loss as fundamental to the nature of reality, just as stasis ultimately denies the reality of evil. Flux encourages, thereby, the metaphysical dualism in modern flux ideologies such as Nazism and Stalinism. Dualism is convenient for totalitarians because it creates scapegoats whose persecution can be used to sanction systematic societal repression. Process, which alone allows for authentic tragedy, sees evil as a challenge to create new good.

From the viewpoint of stasis, time is illusory: the fundamental reality is eternity. Where God lives in eternity, true meaning resides with God outside or above time which encourages escape from fully serious participation in this life and this world. However, in Judeo-Christian views, with exceptions like Spinozism, God can and does act within time. The arbitrary power of Allah, the rigid hegemony over daily life of Sharia law, and the unchallengeable authority of the Koran push Islam further than Judeo-Christianity in the direction of extreme stasis. Like all the major religions, Islam has its variants and inconsistencies which become evident when we juxtapose Sufism and Jihadism.

Flux rejects a consistent, rational benevolent God as creator or unifier of the universe, but modified flux can allow the struggle of good and evil deities among multiple warring entities. Flux can deny that there is any real universe, or that reality lacks fundamental unity, principles, or purpose. All we know is a succession of spurious universes or hypothetical states that appear to the internal observer to be related but are not so far as we can prove. The logical end of flux is atheism. Even so, most flux philosophers will maintain that because the sequential states of time are related, some patterns, if not laws, emerge from the relationships. The apparent flux of the universe would then be unified by the relational nature of time which could allow for limited theism or at least agnosticism. God is, therefore, at least possible in all three views and usually defined, knowingly or not, by the philosophers and theologian's assumptions about time.

While God's existence is at best problematic in a flux universe, for stasis God is real; indeed, God is the ultimate reality. The stasis God creates the universe, but is outside it residing in eternity. (For God to be at once outside of time and the universe, that is in eternity, and the universe's creator and ruler is a paradox, of course.) From the process perspective in this work and elsewhere, Time is the de facto God of the universe that humanity knows. God, being the process of time, unfolds within the universe, and so has no function outside time in eternity. Then eternity, not time is unreal, or perhaps, as we have noted, it stands in relation to time as the zero does to numbers, a place holder without a positive value in/of itself.

Making the process of time pre-eminent over flux and stasis has major implications. Time as a dynamic process is real time. Time as stasis or flux has no reality other than as abstractions that help define real time by showing us what it is not. Humans never perceive either stasis or true flux, we encounter only the knife's edge of process in the present, which becomes another term for real time, passing time, or the succession of events. As long as we are alive, we are immersed in our present. But we have concepts of flux and of stasis that help us define or wrestle with real time in the way that nothingness helps us define or wrestle with being. Indeed, would it be possible for us to understand being without nothingness? Could we grasp conceptually the nature of time as process, which is being, without the facilitating concepts of flux and stasis? Can we understand the present without reference to the past or future? Certainly not, the past and the future are essential tools for coping with the present which contains nothing and everything. Indeed, we cannot think about the present at all, we can only think about the past and the future, but must do so in the present!

Process and its continuum contrast to stasis where universals are set in eternity. Continuum is the fluid succession of moments extending into the past and onto the future, each moment phasing into the next and each imperceptibly different from immediate past and future moments, the whole defined by the process within the continuum of events. The moments themselves are necessary fictions of apperception that lack reality in the noumenon which is pure process. (The question of time as a succession of moments versus a continuum or process resembles the question of whether light is a wave or a particle. The answer is we need to think of them as both. Time is/acts as a succession and a continuum; light is/acts as a wave and a particle.) In this view, processes replace Plato's universals as the structure of knowledge. Organic life, more than mechanical physics, provides the metaphoric keys to our process universe, which in its essential character more resembles life, its ultimate goal, than the inorganic matter from which life initially appears to coalesce. The evolutionary continuum manifests meanings within processes that transcend and nullify the meaninglessness of unstructured flux.

A temporal, process universe is necessarily an interconnected, interdependent universe. The mystery of time is the mystery of interconnection and interdependence: how will events in the present affect the future, how do events in the present reflect the past? In an extreme flux view they do not: everything is independent, all is flux and instead of meaning in processes we have anarchy or at least a lack of essential connectedness. In stasis views dependence on God, the creator, or on the big bang or a mythical event, explains the origin of time and the universe making all flow inevitably from and dependent on the deity and/or the primordial event. That would seem to leave us with a pre-determined universe devoid of free will and creativity. Against such a universe, one can appreciate why forms of Buddhism call for escape from time and its pointless suffering into the eternal nothingness of Nirvana.

Interdependence and interconnection, that which is interdependent is interconnected and conversely, are fundamental assumptions of Jungian thought on religion. They are even more fundamental for Whitehead and process philosophy, and for process theology as well which interprets love in terms of interdependence and respect rather than the dependence and obedience stressed by stasis theology. Jung's prime goal of wholeness entails a harmonious interdependence of psychic functions effected by the homeostasis that the self sustains. Individuation is the process toward wholeness of the interdependent functions in the self. Interdependence, rather than obedience, is the ideal relationship of man to God, and of ego to self as they work to co-create consciousness and novel value in the world.

Jung's Quaternity symbolizes the interdependence of the four archetypes that constitute the psychological reality of the Christian deity. Trinity represents not the entire deity but the three-part dialectical process moving toward the divine individuation symbolized by the Quaternity. Quaternity entails metaphysical uniplurality, a theology of process, and an ethic of interdependence. Quaternity and Trinity reflect, for Jung, the interdependence of past, present, and future in the dialectical process of the universe that is Time as God.

Without interdependence and interconnection nothing could be predicted because nothing would be related to anything else. Thus, the universal interconnection created by and characteristic of time makes science as well as comedy, tragedy, and religious meaning possible. Without interconnection there can be no tragic loss, comic gain, religious meaning, or scientific knowledge. You cannot lose, gain, or predict anything about that with which you are not interconnected.

We are interconnected with all, yet degrees and qualities of interconnection vary. Variance is especially true in respect to temporal proximity. However, temporal proximity is only one factor. Our interdependence with historical figures, say Christ, Mohammad, Lincoln, Hitler, Newton or Darwin, may be greater than with others sharing our own timeframe because their influence fields may extend much further.

So Mohammad or Darwin, who have become time beacons, may have far more impact on us than Mrs. Smith who is our contemporary living three houses down the block.

To summarize, philosophers and other major thinkers fall into a spectrum where their basic positions relate to whether they see time more as a stasis, a process, or a flux. This work, along with Whitehead, Bergson, Jung, Shakespeare, and many others, assumes that time is process. Process makes interdependence, rather than dependence or independence, the fundamental feature of reality. We all subsist in complex, interdependent ways within our contexts or tardises in the processes of time. Time requires the interdependence and interconnection of the succession of events and moments. Interdependence and interconnection sustain succession and continuum and make time the supreme relational process and universal regulatory system. But interdependence is ultimately unreal in the stasis and flux views which assume respectively either absolute dependence on God or universal independence. The meaning of time in the present, as well as in the overall process of time, is defined against flux and stasis. Flux and stasis are meaningless in themselves because they exclude interdependence which is necessary for relationships and, thus, meaning and value which always entail connectedness. But flux and stasis help define the interdependent process, the most rewarding itinerary for the philosophic quest.

MEDITATION XII: IMAGINATION: EVERYDAY TIME TRAVEL

Imagination is a unique and essential activity of consciousness. A bigger question than "can we develop machines that think?" is "can we develop machines that imagine?" Much of thinking requires imagining, logical thought is but one aspect of human thinking. Time travel is not only possible through imagination; time travel is a mundane activity of the imagination. Imagination is how we connect with the past and the future. We re-imagine the imagery from our past, trying to grasp or envision what happened as it happened. Through imagination we attempt to recall and understand the context, causes, and consequences of a single event or a subsequent course of events. We speculate on how a different choice or decision, might have changed the direction of our lives and the lives of others, or we simply try to relive critical experiences or relish intense experiences. We also try to imagine, to envision, the future and make our present decisions on the basis of what we imagine our future can be or what our present decisions impact on the future will be.

It is no wonder that man speculates about actual time travel when he does it all the time in his dreams and in his imagination through fantasies. In our dreams we travel spontaneously into a re-imagined past or an imagined future we expect, hope for, or fear. Our dreams exist outside of time in that their sequences do not obey linear time or normal causality. Oblivious to clocks, the psyche arranges events in dreams by associations of affect, imagery, themes, and archetypal energies. Science fiction focuses on literal, physical time travel, as when we fly in an airplane from Washington to San Francisco. These mundane, literalistic fantasies are only one and by no means the most intriguing of the myriad possibilities of time travel. Imagination can open rabbit holes for travel into a universe of temporal worlds, or tardises, like but significantly different from our own.

We live our lives like passengers on an express train. First moving rapidly thru a dark tunnel then breaking into light at birth. Adding new cars filled with different people, each stop inaugurating a new life stage. Plunging boldly forward at first, but still adding cars until our memory train's very length and mass begin to slow it down. As the train grows longer, cars at the rear fade and darken one by one. We must always live nearest the engine, moving into each new car as it is added just behind the engine which is our life force propelling consciousness into the light of the future. But as we age the train slows from the mass of accumulating memories, it's dusk world turning to night, until it stops, all movement within freezes, and the light of consciousness, once so brilliant, is extinguished. Each train car is a tardis with its own peculiar qualities, people, and active archetypes. A few people move forward with us into the new cars with the new people joining our lives; but most stay behind converging into fixed portraits of our past. Some travelers will try to cling to the car they are in and may do so for a while, but events always propel them toward each newly created car behind the engine. Train wrecks, early deaths, come to some. Those who cannot move forward gracefully into the next stage are more prone to train wreck. How can they guide their trains remaining too far from the engine of time? Others crash thru sheer ill luck not of their own making, but all appear to end in immobile darkness from the perspective of those still alive, still moving in their own trains.

Assuming the prevalent Western conceptions of time based on science and technology, as well as rationalist philosophy, physical time travel would seem to be impossible. We could not go back to the Ford Theater in 1865 and stay the hand of John Wilkes Booth or to Sarajevo in 1914 and foil the plot of Gavrilo Princip, as conducive as these actions might be to subsequent human welfare. In orthodox Judeo-Christian theology, God sets the rules of the universe from creation, which is a boon to the pursuit of science, if not to physical time travel. He cannot, then, change them, because these rules reflect His essential character and vouchsafe its continued goodness and rationality. God is truth (rational) and God is love, that is the Western God's very DNA. As Aquinas stated, God cannot change logic or geometry or make evil acts good. The range of possibilities under Islam's Allah, however, is unlimited because he possesses a totally arbitrary power that knows no bounds or restraints. The mind of Allah presumably could change history by making Calpurnia screw her courage to the sticking place, keep Caesar at home with her in their palace on the Ides of March, thereby forcing Dante's Satan to find a different flavor of chewing gum.

The irreversibility of time is an unacknowledged absolute in Western culture. This is evidenced by our making the Son of God's earthly life, suffering, and crucifixion our paramount myth, tragedy our highest art, and *King Lear* the supreme tragedy. In that framework, a non-physical form of time travel might still be possible. If in our minds eye of consciousness we could journey back to view events without acting in them, the fundamental nature of time, it's irreversibility, might remain intact. We might then see exactly where Brutus stabbed Caesar and note the expressions on the face of the murderer and his victim. We might even return to paint the expressions on the faces of Caesar and his assassins. But, we could not return with a photograph, only with a memory. Our journey would be mental, not physical. Such a feat would require consciousness to disembody from the material body, which in turn would mean that the psychic and the material must be separate, though interconnected, realms. Such disembodiment of mind or spirit has been a source of speculation since the dawn of history. Simple people throughout the world believe in ghosts and hauntings: these phenomena appear to require a separation of consciousness from body. Their beliefs may evidence what phenomenologists, like Husserl and Merleau-Ponty, might term pre-reflective knowledge, in this case knowledge of the separation of mind and psyche from body.

Forms of time travel might be facilitated by time fields or tardises. Information from one tardis might become accessible through an archetypal link to a related tardis. The tardises might actually have an organization in some respects akin to biological organisms and possess limited capacity to communicate or even interact in a type of community, as some speculate that plants do. Thus, tardises might interact causing, or at least allowing, forms of precognition or retro cognition. Consciousness, or mind, as thought, feeling, intuition, or even perception might then enjoy a mobility in time that is not possible for physical objects.

History writing too, arising as it does out of story-telling, is a conscious endeavor at time travel, an attempt to reconstruct the events and meanings of a period in the past. History does not resort to the supernatural or violate our scientific understanding of time. Nonetheless, traveling to the past, either within alternate forms of consciousness, or by story-telling and historical imagination, creates an additional means beyond ordinary causation whereby the past can alter the present and the future. We could not travel into the past even in these non-physical, mental ways without altering ourselves, for we are the sum of our connections and relationships which expand through the exercise of memory and historical imagination. Thus, memory, even the more limited memory faculties of creatures with smaller brains than homo sapiens, becomes creative by allowing the past to alter our present and direct our future in new and unexpected ways. So dynamic consciousness mediates between past, present, and future to make our temporal realms like interactive and interconnected organisms. In this respect, consciousness may transcend time to become in itself the supreme creation of the temporal process. Ultimate consciousness transcending time, many theologians and philosophers believe, is an essential attribute of God

.

MEDITATION XIII:
TIME, CREATOR OF VALUE AND BEAUTY

In a temporal universe all things come into being and exist for a time; yet given enough time all are irretrievably lost. Time cycles of creation and destruction give rise to value which, much like consciousness, resembles an epiphenomenon or, if you will, a psychic phenomenon. (Jung, it should be noted, believed consciousness is fundamental and rejected the term epiphenomenon, but this seems a semantic issue.) With both value and consciousness, the "epiphenomenon" is what gives meaning to the material process and all that happens therein. Indeed, value creation may be seen as the de facto purpose or teleology of Time and of our time bound universe itself—a purpose lacking an identifiable, ego-like, purposer, or at least one whose nature and character all can agree on.

God operates not like a decision making ego, but like a homeostatic self, Jung surmised. The paradigm shift from ego to self is Jung's Copernican revolution in psychology and theology. These disciplines, along with cosmology, have yet to incorporate the paradigm shift, because the crucial difference between ego and self is not yet widely recognized or understood. We are only beginning to grasp its vast net of implications. Chief among these implications is that Actual God, the homeostatic self of the cosmos manifest in time, the creator and guarantor of value and meaning, is essentially different from the monotheistic and polytheistic gods nearly all mankind worship which are powered by activated archetypes and modeled on the ego and/or related subordinate psychic functions such as the anima and shadow. The god image in the psyche reflects not these subordinate functions, but the self itself which is a microcosm reflecting Actual God, the ordering force of the macrocosm or the cosmos.

A universe without time would be without value because there would be no distinction between any part and another, and no consciousness or separate centers of consciousness, e.g. minds, to distinguish and evaluate anything. Value, good, evil etc., depend on temporality and cannot exist without it. Those who argue that "God exists" are likely arguing that we live in a universe where value exists. That value exists is a given entailed by the temporal process, the *creative continua*, of the universe which *continuously creates novelty because it makes all things and events unique by giving each a singular position in time and space where all are uniquely interconnected, yet fleeting.* But the religious do not usually mean that value exists when they say God exists. They mean that their own ego-based conceptions of good prevail universally because they are upheld by an omnipotent egolike deity. Their belief stems from faith, hope, and limited perspectives, not from understand the nature of the self, time, and deity.

Traditional theologians argue that God is pure good, the *Summum Bonum*. As Jung and Whitehead stressed, this view creates insurmountable problems if they also insist that God is omnipotent, omniscient, and providential, and therefore both able and inclined to arbitrarily intervene in temporal events to insure specific outcomes favorable to individual sentient beings. If God is in truth the omni-omni *Summun Bonum*, there should be no suffering or irreparable loss and no need for divine intervention. Tragedy dramatizes that men do not live in a universe ruled by such a god. The theodicy problem recedes if we stop claiming that God is the *Summum Bonum*, but instead say that God is the process of unfolding the self of the universe through time. God unfolds thru time, as does the good, the true, and the valuable. Without time, the driving force for God and consciousness, the good could not exist, or subsist as the epiphenomenon we perceive as value.

There is not necessarily a beginning or an end to time itself or the universe: change and transformation are constant. Was the universe created out of nothing? That may not seem to make sense, but neither does creation out of something totally unknowable which may amount to the same thing as creation out of nothing. (Since this is the only universe we've got; we can't compare it to anything else.) We do not know what brought the material universe into existence, what if anything preceded the imputed "big bang," let alone why matter is paired with mind. Things in time appear to have a beginning and an end, but these are just milestones in the ever unfolding processes of the continuum of time. Like the rail cars in the lengthening train of life, beginnings and ends are mental constructs that help the mind find a way to order events--or to hold at bay the ego's intolerable fears of being lost and alone. The order may be imposed by mind, and may not be a characteristic of external reality itself.

Time may well flow like a river with no discrete breaks. It may be an unfolding process or continuum with no known, and possibly no knowable beginning or end. Events may be just constructs of the mind, but that would not make them meaningless or without value. The universal flow of time, some speculate, might be a circle, a uroboros of the interdependent, interactive energies of yin and yang. Or the flow may be likened to a creative process that begins in a vast mystery and slowly spirals off into space, a process with no easily understandable beginning or end. If it does not end, or begin, can the universe itself have value? Not so, value refers to the events and things with beginnings and endings within the time process that comprises the universe. The universe is a context for value, like the frame of a picture. Time, the *creative continua*, is the creator and sustainer of value and of the universe itself.

A mystery: God, or whatever you call the force that brought the universe into being and sustains it, brought time into being to create value and meaning, and to manifest truth and beauty. (God said let there be time and there was value, meaning, truth, love, and beauty.) Beauty is the mark of value and meaning, and of some truths as well. Look about our world. You can see beauty in its endless forms unfolding everywhere. One could argue that this beauty is in the eye of the beholder, but the beholder cannot see beauty where there is no beholder and nothing to behold. Of course the beauty is in the eye of the beholder, beauty is a value perception requiring a conscious perceiver capable of feeling. Yet that does not make the beauty any less real. The universe is bound in time and beauty unfolds thru time.

Beauty thrives in the variegated processes and surprise filled flux of time. Stasis is an appearance, a hopeful, convenient construct of the mind, never a permanent reality. The Greeks, enamored of stasis, sought in changeless perfection their ideal of beauty. It took the challenges of Christianity and Islam to break the sterile hold of rigid Hellenic aesthetic ideals and open new realms of living beauty in painting, poetry, and drama, along with magnificent architectural, ceramic, and textile forms. Beauty cannot remain static except in our mind's eye. In the world it is always fleeting, or in process. Flowers always fade and die. Their process is their enduring reality and beauty.

So what can we say of the Actual God who created the process that unfolds beauty, value, truth, and Itself in such a universe? That God is not preeminently a moralist, however convenient that conceit might be for those who earn their living by preaching lifeless obedience to mind deadening social and political conventions. Tragedy, above all *King Lear*, shows us why and how it becomes impossible to believe in God the rigorous and consistent moralist, the *summum bonum*, against all the suffering, waste, and sheer brutality and absurdity we know and must ever confront. If God cares about morality, it must be moral beauty, the beauty of wholeness, authenticity, truth, integrity, love, and joy.

No moralist, Actual God is an artist, a belief apparently held by Whitehead and assumed by Shakespeare and many great artists. The universe is God's defining art work, the supreme epic, and time is God's medium. Beauty is the meaning of the universe and of life. God's purpose in creating the universe cannot be to insure the happiness of man or any beings, beyond the fleeting happiness afforded by beauty, its creation and unfolding. Individual happiness is born of participation in the beauty that is created or unfolded within every being and each thing around us all the time. (A lesson from Shakespeare: Love, compassion, and truth are exalted forms of beauty.) Youth has its wisdom, it special insights just like old age, as twenty-three year old John Keats sometimes demonstrated:

> When old age shall this generation waste,
> Thou shalt remain, in midst of other woe
> Than ours, a friend to man, to whom thou say'st,
> "Beauty is truth, truth beauty,—that is all
> Ye know on earth, and all ye need to know."
> *Ode on a Grecian Urn*

But age has superior judgment and can make finer distinctions. Whitehead was asked: "Beauty is a wider and more fundamental notion than truth?" He replied, "Yes, apart from beauty, truth is neither good nor bad." Beauty is often the truest measure of value.

"What is truth?" Pontius Pilate famously asked. An equally difficult and important question is, "What is beauty?" A string of adjectives, some expected some not, apply to beauty. Beauty is harmonious, but it is also surprising and both symmetrical and chiral. Human faces, for example, are chiral yet must retain basic symmetry to be beautiful. St. Augustine said that beauty displayed wholeness and Jung would doubtless agree; integrity and unity are characteristic of what we deem beautiful. Still something broken or injured can be beautiful, because beauty must arrest the eye and can stir compassion as well as desire. Beauty is enjoyable, yet the beauty in tragedy and our struggles against evil can also be excruciatingly painful, but never meaningless. Beauty surprises us to gain and hold our attention, and it stirs a sense of wonder to make us think. Beauty is magical: there's magic in the web of it, as Othello remarked of his strawberry handkerchief. Sublime beauty evokes awe. Part of beauty's surprise is its comparative rarity. Still things that are not rare can be differently beautiful.

The Sun's planets are all very different and beautiful; like the planets, men, women, animals, and plants can be differently beautiful. Some things may not seem beautiful in themselves, small asteroids, grains of sand, empty space, dull people, mushrooms, sea cows, hyenas, but even they acquire beauty by their participation in larger wholes, in their unique roles in the unfolding web of life, the universe, and God. Jane Austen's character Miss Bates seems dull and lacks beauty in herself, but she is a crucial element in the beautiful story her creating artist unfolds. The compassion she merits and elicits in others is her essential beauty. Emotions can be beautiful or ugly, and both forms of emotion can be elicited by ugly as well as beautiful things.

Beauty can inspire great thoughts and actions, soothe fear and anger, and heal suffering and despair. The beauty of a waterfall, a mountain, or a starry night can do these things while evoking thoughts of God. Christian thinkers saw that God is love and truth. The great Christian artists knew feelingly that God is also beauty, that we know and worship God through creation of beauty as well as through love and truth. Indeed, love and truth always have their own beauty; and beauty has its own truth. As Plato shows in the *Symposium*, beauty inspires love that can lead us to truth. Plato and Aristotle recognized that the artist's inspiration to create beauty has a divine source. Nearly two millennia of Christian artists show the countless ways beauty leads us back to its divine source.

MEDITATION XIV: SIZING UP OUR GODS

God, as conceived by most Western monotheists, resembles a human personality with an ego, a temperament, and a personal history. God has values, preferences and even idiosyncrasies, makes decisions, and sets rules through a mind guided by emotions that are often reassuringly or disturbingly similar to those of a human being. Such a deity is not that difficult to make sense of when placed in the context of the tiny geocentric universe of Moses, Jesus, Hillel, Augustine, Mohammad, or Aquinas. But a God like a human ego became increasingly less credible as our knowledge of the universe expanded from geocentricity to a solar system, and on into a galaxy where our Sun is one of 400,000,000,000 stars, where possibly 3 billion planets might support some forms of life, all in a universe with 7 times as many galaxies as our Milky Way has stars, and some of those galaxies and stars much older than ours, with billions of years more time to develop advanced life given that the universe appears by current scientific estimates to be approximately 13,772,000,000, not 6000, years old.

There is an optimum size for most things, stars can remain stable at as much as 100 times the mass of our sun, but not 1,000 times or they will change into something altogether different, like a supernova or a black hole. Animals the size of whales, or the 110 ton Argentinosaurus can still function, but there have never been animals anywhere near the size a 250,000 ton cruise ship or even a plant the size of the Pentagon or a tree as lofty as the Burj Kalifa. Assuming that consciousness is limited by the speed of light like everything else, given the scope of the universe, the vast extent of time and space, it is impossible to conceive rationally how a single god with one personality, ego, and mind, could "rule" the Milky Way galaxy, let alone the entire universe.

Such a concept of deity approaches credibility only if we assume a geocentric universe in a time and spaceframe an infinitesimal fraction of the size of what science has revealed. The very nature and extent of time and space in this universe render inconceivable an anthropomorphic deity like those of our legacy monotheistic religions. The omnipotent, omniscient, all beneficent, ego-like monarch deity of standard Islamic, Judaic, and Christian theology simply defies logic and known scientific principles and facts. The cultural dominance of these religions across half of our tiny, flyspeck world is not a plausible argument for their moral authority, let alone their veracity.

Many other notions of God remain to be explored, and some of them may be plausible in reference to modern conceptions of the Universe. Thus, the short comings of the human monarch model of deity are not valid arguments for atheism even though they are naively and also cynically exploited for that purpose. New understandings of deity, as in Jungian psychology or Whitehead's philosophy, or ancient concepts like Taoism, may prove meaningful in the time-space frame of modern science.

However, if Actual God is perfect unchanging goodness, omniscient, omnipotent etc., It-He-She cannot be within the universe because the universe is a temporal process where all is subject to change, and change is often unpredictable. Even basic particles decay over time, nothing can be perfect. If It-He-She is outside the universe, then It-He-She cannot impact the universe without becoming part of it and thus subject to time and change. So a perfect God of the Universe who, like Spinoza's god, is wholly independent of our universe would be entirely theoretical and have no existential meaning, impact, or value. Viable conceptions of Actual God must necessarily be pantheistic, that is make the deity immanent and interdependent with the universe, as well as subject to time and change like all within the universe. Otherwise Actual God does not exist in any sense that can be relevant to any conscious beings within the universe.

Actual God does indisputably exist if defined as the process of time, change and value creation, as the ultimate origin and *raison d'etre* of Value, Meaning, Truth, Love, and Beauty. Man, at least at the current stage of human development, has difficulty determining whether the process might prove to be scientifically intelligible, or if it is shaped by some form of intelligence so advanced or different from ours that it is and will remain in important respects beyond human comprehension. But why depend on science to answer questions of value and meaning when these come from emotion guided by intuition and imagination? Myth is the product of emotion, intuition, and imagination. As long as myth is not corrupted by ego fixations and prideful rationalism, but reflects authentic human feeling and experience, myth can provide a ready and plausible path to understanding a form of deity that exists and acts within the temporal process.

Jung, or Whitehead, might say that Actual God is intelligent or conscious not in the sense that an ego is, but instead possesses a process intelligence more complex and multifaceted, and more broadly focused, like that of the homeostatic self. Such a process God could be intuitively manifest in the big dreams of our culture, our paramount myths such as *The Book of J*ob, the Gospels, and *King Lear*, even as the self is manifest in the personal dreams of individuals. Such a God would not be oblivious to the human condition or alien to the human heart.

MEDITATION XV:
THE HUMAN CONDITION

1.TRAGEDY. More important than the fated tragic end is the tragic struggle to determine who we are and shall be by asserting our integrity against the ravages of time. Consider *King Lear, Moby Dick*, *The Duchess of Malfi* and two of cinema's most admired films, *Sunset Boulevard* and *The Music Room.* Each has a splendid protagonist struggling to maintain her/his integrity against time, against the cruel indifference of nature and society. We see in bold, dramatic strokes the pathos of life's struggles with time the destroyer, the deceiver, the betrayer, the utterly callous. Lear, Captain Ahab, the Duchess of Malfi, Norma Desmond and Satyajit Ray's Maharajah each fall before time's irresistible sword. But the pathos of their struggles, and their courageous affirmations of their individual worth and defiant defense of their integrity, create tragic dignity, moral courage, and spiritual beauty that time cannot tarnish or destroy. In the words of John Webster:

> Nature does nothing so great for great men
> As when she's pleased to make them lords of truth.
> *Integrity of life is time's best friend,*

Which nobly beyond death, shall crown the end.
The Duchess of Malfi, V.v.

Webster's integrity of life is a version, presented within a tragedy, of Shakespeare's vision in the Sonnets of art staying time's cruel hand in a moral triumph over time the destroyer. Thereby, integrity can make an individual life a form of art proof against time. Most Elizabethan tragic heroes seek to maintain the dreams of youth, the springtime when the sheer beauty, will, and the energy of life seems able to hold time the ravager at bay. Prince Hal believed it would, while Hamlet despaired at his recognition that the dreams of his youth must soon end in tragedy. We identify with their hope and heroism even though it may leave us uncomfortable by setting too high a bar. The impact of their heroism makes them sympathetic, like real suffering individuals whom we have known, rather than mere figments of an artist's elaborated fantasy. Indeed, we feel we know these tragic heroes because we know and feel their angst of time all too well. They do not adjust to time's cruel hand; they refuse to go gently into their good night. They defy and valiantly battle time the destroyer's relentless drive to corrupt integrity by defining our identity downward into something less than what the self can realize.

When time offers opportunities to define ourselves upward as the wheel of fortune turns up, fools indulge inflation or hubris, a deceptive type of identity change whose effects can be as tragic as identity loss. However, there is a third way of handling identity change that is not reactive, like inflation and deflation, but instead creative. Through the process of individuation, wise souls can overcome time by unfolding within it: for example, Shakespeare's Prospero or even tragic characters, like Hamlet, Lear, and the Duchess of Malfi, who realize a qualified triumph over the tempests of time. Where the triumph of individuation is unqualified, and that is seldom and fleeting, the product is comedy. As the great actor Edwin Booth observed, comedy is difficult. To unfold or realize a human comedy of individuation in a person's life is very rare and difficult indeed.

In the West the stasis view leads toward absolutism which usually stops short of that final conclusion as Western religions introduce the dogma of man's free will to exempt their God from responsibility for evil in his creation, and to mitigate the tragic aspects of time and human life. Tragedy has unique importance in the West, as is evident by our deeming it the highest form of drama and the apex of our literature. Tragedy stands apart as a cultural qualification and challenge to religious orthodoxy, and to stasis philosophy's claims for an omnipotent, omniscient *summum bonum*. Since tragedy is fundamental to the human condition, the concept and experience are universally understood even though tragedy is a quintessentially Western art form. The story of Christ is a kind of tragedy, especially as re-interpreted by Jung setting aside St. Paul to explore the roots of the crucifixion story in Job's unjust suffering. Thereby, Jung laid bare the archetypal core of Judeo-Christian myth. That raw core, as I have often stated, is most clearly exposed in the narrative, symbolism, and imagery of the greatest of all tragedies, *King Lear*.

2. TEMPORAL BLACK HOLES. Death in a pure tragedy like *King Lear* or *Macbeth* can be seen as an ultimate temporal black hole, and

dying would, then, be like crossing the event horizon. It represents irreparable loss of identity, the great dread that Lear, Captain Ahab, and Norma Desmond share (the Duchess of Malfi and Christ are different stories). Christian and Islamic orthodoxy reject this tragic view of time. For orthodox Christianity and Islam the soul is eternal, the black hole of tragedy cannot swallow it. Both religions have a last judgment that marks the end of history opening into a comedic era of endless happiness for the believers—mansions for the Christians and forty virgins for the Moslems! There are variations where non-believers get purgatory, but unrepentant sinners are forever confined to the torments of hell. It should be noted that in Dante, Milton, and other explicitly Christian writers Hell is not a black hole of lost identity. Consciousness persists as a separation from goodness filled with feelings of eternal loss, regret, and suffering. The tormenting memories of lost identity and lost hope never fade: that is the fate of the damned.

Western tragedy began with Greek drama and arguably achieved its artistic zenith in Shakespeare with honorable mention going to *The Duchess of Malfi,* Milton's *Samson Agonistes*, Herman Melville's epic novelistic adaption of elements of *King Lear* in *Moby Dick,* and to Dostoevsky's *The Brothers Karamazov* and Tolstoy's *Anna Karenina.* In the East re-incarnation means no temporal black hole for individuals, only the purgatorial cycles of birth, life, death, and rebirth in a different form. Some aspects of identity are not lost in death. The soul continues, for it is indestructible, and even the conscious self may be preserved in a living person's "past life" that can be recalled in fragments. Nirvana might seem an Eastern version of the temporal black holes that define Western tragedy.

However, that might not be so if the individual self or soul melts into a universal soul—which the West might interpret as God. But for those who follow Macbeth and see life and death as a tale full of sound and fury told by an idiot, the black hole of death signifies the nothingness envisioned by atheism.

3. TRAGEDY in HISTORY. Tragedy focuses on and arises from a disastrous break in the harmonious flow of time. History affords tragedy aplenty, with extraordinary tragic heroes and villains to match. Jesus the man's story resembles a tragedy. Though the perfect Jesus of orthodox theology cannot be a genuine tragic hero as the Son in *Paradise Lost* shows. According to Jung's later day theological revelation, *Answer to Job,* the Father, or Yahweh, is the real and flawed tragic hero of Biblical myth.

The supreme tragic hero in the last millennium of Western history is an American president, Abraham Lincoln who like Hamlet is caught in a time out of joint. Like Hamlet he was noble, unforgettably articulate, and anguished by being forced to right wrongs not of his making. Lincoln may provide the consummate window on historic tragedy because we know so much about him. His facts are far richer than with Job, Socrates, Jesus, or the fictional Hamlet. Lincoln's story, however, remains the raw material of tragedy because Lincoln has no Shakespeare to tell his tale, only historians who fall short in poetic and philosophic genius. So why is Lincoln tragic, why is he not just plainly heroic like George Washington? Here I must enter the dubious plain of historical speculation.

Certainly, Lincoln's death was a tragedy for him and for the nation he led. But there is more to his tragedy than just a violent death through no fault of his own. The Civil War is America's equivalent of the Trojan war, a culture wide wound, an ever flowing font, and source of civilization defining myths and values. Seldom has a single war so shaped a great nation and its civilization. At its center stands one towering leader without whom all would have turned out differently—and worse. Abraham Lincoln is our preeminent real, rather than fictional, tragic hero. Great as he was as a human being and leader, he failed to steer the ship of his nation undamaged through the storm to a calm, safe harbor. Although he preserved the union, he failed to save the nation from its divisions against itself. That is Lincoln's tragedy. Ever well intended, a model of character, wisdom, and compassion, the results of his work did not go seamlessly well.

Nearly a fifth of America's fighting age males perished on the battlefield or from war inflicted diseases and wounds. Millions fled their homes depopulating the old Confederacy. Slavery soon yielded to KKK terrorism and Jim Crowe oppression. Corruption metastasized throughout society in the Civil War's wake. One hundred years later a still profoundly divided nation endured the Watts riots of 1965, then in 1967 the Detroit riots. Finally in 1968 the assassinations of Robert Kennedy and Martin Luther King formed a tragic riposte to that of Lincoln. Lincoln himself failed to forestall the Civil War, failed to negotiate a peace to stop the war once it started, failed initially to find generals who could end it quickly, and failed to lay out plans to heal and rebuild civil society once the war was won. He was a great spirit like Churchill, and like him he had a way with words. One wonders what Shakespeare might have done with Abraham Lincoln. Lincoln himself gave us a taste in his Gettysburg Address.

Here is the key to Lincoln's uniquely powerful charisma: more than any major historical figure, Lincoln stepped credibly into the Christ archetype. My surmise is that the historic Jesus was in some ways rather like Lincoln, and Lincoln sensed the similarities and modeled himself unconsciously. The similarities were crucial because America in the 1860s was a profoundly troubled but deeply Christian nation, longing for a good shepherd. We needed someone to remind us that human dignity could survive and greatness emerge from the unspeakable squalor, horror, and anguish of our crucifying dilemma, slavery. America was a nation brought up on the King James Bible, Milton, and Shakespeare. Lincoln drew on all of these and his own humble beginnings to give his nation hope that, despite the continual failings of humanity, noble heroism and sacrifice can redeem us, somehow, from ourselves.

Lincoln was a tragic hero, a great man who made mistakes. Like all tragic heroes driven by ambitions, he lacked the wisdom he needed to unfailingly guide ambition toward the good of all. Yet through greatness of spirit and personal sacrifice he redeemed himself from his human flaws and restored the nation he led. Lincoln did not save us from disaster, but his heroism and sacrifice did help redeem us from our most deadly collective sin, the institution of slavery.

Tragedy gives value to human life, rescues us from emptiness, despair, and self-loathing. Only blind fools would dismiss as futile the struggles in the stories of Jesus, Lear, and Abraham Lincoln. Must not they tell us something about time and the human condition, about the great suffering God of the West? Tragedy gives meaning to time's irreparable losses, and it helps redeem our humanity by allowing us to transcend suffering and evil through creation of truth and beauty. Such is tragedy's lesson, a lesson that makes it the quintessential expression of the spiritual genius of the West and the power of the archetypes of its Godhead.

4. TIME AND TRAGIC ADDICTION. "And I was very happy, for a time." – Listen to this climacteric line from *Long Day's Journey into Night* by Eugene O'Neil, as spoken as only she could by the great Kate Hepburn. We all have our addictions, our attempts to regain, sustain, and extend memories of fleeting happiness. Addiction is an essential feature of our human nature and condition, bound as they are by time, tragedy, and death. It's one of our many different attempts to re-enliven precious moments of happiness and meaning that allow us to escape, if only for a while, time the destroyer. The characters in O'Neil's play, like so many in the dramas that make up our "real" lives, cannot cope with defeat, loss, and time unfolding. They long for stasis, eternity, and they fear change and flux. Unable to cope with time as process, they become addicted to precious moments, not realizing that in the moment there is nothing. For all meaning comes through engagement in the entire time process rooted in the past and flowering into the future. The audience's recognition of the great truths about life's temporality is the gift of tragedy, from Sophocles through Shakespeare to O'Neil: a gift made possible by the

sacrifice of the tragic hero. That sacrifice has become iconic for our civilization with the characters of Oedipus, Lear, Hamlet, Othello, Macbeth, Captain Ahab, and above all the historical figures of Socrates, Jesus, and Lincoln. Their tragedies are the result of sublime and fatal passions, obsessions, addictions, and commitments contending against time the destroyer and the cruel limitations it imposes upon humanity and all life. Their sacrifice gives catharsis, liberating us from enslavement to obsession, addiction, and, above all, the enervating, soul sapping mundane.

5: ALL LIFE CONTENDS WITH TIME. Conscious human individuals form modes of apperception that constitute adaptations to the temporal realities of our world. Animals and plants do also, but human adaptations to time are more conscious, and, because of our technology, more complex and broadly consequential. Traditional philosophers predating Darwin and modern biological science, including Kant and Hegel whose influence on post Darwinian philosophers like Marx and Nietzsche was great, tend to see an absolute separation between human intelligence and the intelligence implicit in evolving life itself. For them man alone possesses a mind, a spirit, and a soul. Descartes, an extreme human chauvinist, dismissed animals as mere mechanisms. But modern biological science shows us that animals and other life, such as plants, are no more mechanisms than is man, and they are usually as attuned to time changes as we are. DNA studies reveal that humans share 99% of our genes with Chimpanzees, and with cats 90% of our genes are the same. Primitive forms of life, no less than advanced, have special sensitivities and complex adaptations to time that machines and lifeless matter lack.

All life is driven to survive, adapt, and reproduce itself in contention with time. Simple celled organisms and humans alike are driven to maintain and extend the cellular order and propagate the DNA that sustains their identity. Higher animals have a strong sense of their individual identities and their boundaries. In an existential sense, all life is at war with time in its struggle to maintain identity, while inert matter is at peace with time. Life, through its contention with time, its drive to maintain, extend, refine, and reproduce identity, creates the apperception structures and functions that order time in an attempt to conquer, or at least hold time at bay in order to manage it. Time always wins against individual living entities in that it extinguishes their identity in death. However, it's an open question whether life itself can ultimately win imposing its value on all time, thus turning time and the universe it creates into a living thing whose meaning becomes life itself. Would life be, then, the acorns of God unfolding in the universe?

MEDITATION XVI: GODHEAD TODAY

Peoples constellate archetypes to form gods within contexts set by historic time fields. Whether monotheistic or polytheistic, a civilization's god or gods form a godhead whose characteristics reflect the civilization's common aspirations, hopes, fears, biases, traditions, and values. The figures of Father, Son, and Holy Ghost compose the orthodox Christian Godhead. Brahma, Vishnu, and Shiva form the Hindu godhead; Zeus, Hera, Hades, Poseidon are the chief figures of the polytheistic Greek godhead. The Islamic godhead is singular: Allah almighty, Allah alone.

Godhead refers to the essential nature, character, and structure of the deity or the complex of divine figures that shape and dominate a culture or civilization. Godhead forms through a coalescing of traditions and ideas around dynamic archetypes. The formation of a Godhead mirrors a fundamental coalescing process in life and in the universe wherein disparate parts gather and combine into a novel whole. The coalescing process is evident across the cosmos. Stars coalesce into galaxies, interstellar matter coalesces into stars and planets. Subatomic particles bond to form the electrons, protons and neutrons in atoms, chemical elements coalesce into a wide range of complex compounds etc. In the microcosm of humanity, families coalesce into bands and clans which become tribes which coalesce into nations, and nations coalesce into vast empires, like Rome or China, that become civilizations that often form their own Godheads.

Within the psyche, Jung maintained, the self, in response to the challenges of temporal existence, guides and impels its various functions to coalesce into a distinct person aware of his/her social and personal identity. It then moves that person-ego to seek wholeness. In like manner, the dominant ideas, values, and archetypes of a culture or civilization coalesce into a Godhead. The civilizational Godhead assumes a homeostatic function like that of the self in a person. It must be stressed that the archetypes constellating a Godhead never coincide perfectly with official images of them in the cultural canon: the archetypes are independent psychic dynamisms. Indeed, the disparities can become creative forces driving growth and change, as they frequently have been in the Christian West.

Whether or not, or how far and in what ways, the common people believe in a specific distinctive divine being or beings, a highly developed civilization will have something that acts as a Godhead. If the symbols and rituals around conscious worship of the traditional Godhead are repressed politically, a shadow Godhead may form to reflect the dominant values and dogmas of the ruling political elite. Despite the Soviet Union's official dogma of atheism, the ideas and images of Marx-Lenin-Stalin operated like a Godhead in that culture and society.

A civilizational Godhead is the root and ultimate touchstone for that civilization's core values and identity. The worldwide decline of religion among our secular intelligentsia causes disinterest in Godhead and in the quest for religious truth. In more religious ages, like the time of St. Augustine on to that of John Milton, a controlling purpose, a Holy Grail quest, among educated religious people, was often, explicitly or implicitly, their search for truth about Godhead. Such truth presumably brought you to God which meant it could save your soul—at least it meant that for persons of high ability and distinct religious character like St. Augustine.

Today religion is rarely a wholehearted quest for an evolving truth of Godhead manifesting/ creating itself in history. Modern people lack intellectual seriousness about religion. They grab their groupthink from their culture's stale, dried out smorgasbord of stock ideas. Even those of high intellectual ability seldom bother to probe doctrinal formulas in quest for religious truth and/or Godhead. Their underlying, if unacknowledged, assumption is the unattainability of truth. Most modern "religious people" are agnostics or uncritical group thinkers posing as religious devotees. They are Churchians first and foremost, and Christians later if at all. Others who reject religion and God outright become equally uncritical. Failure to seek religious truth weakens a culture. Rather than creative coalescence, the culture falls into its opposite, entropy. The decline of the Russia is a prime example of cultural entropy. When Yeats wrote "*Things fall apart; the centre cannot hold; Mere anarchy is loosed upon the world,*" he could have been describing equally the fall of Czarist Russia and of the Soviet Union decades later. How different Stalinist Russia was and Putin's Russia is from the Russia of *The Brothers Karamazov*.

Jung endeavored to show that, regardless of the failings of one's culture, the self impels everyone on a personal quest for truth which at bottom is a religious quest. The God image is present in every psyche; we can no more rid ourselves of God than we can rid ourselves of a self. The self seeks and can find truth and wholeness for the individual in any culture no matter how decadent, but the task is much harder in some than in others. Individuation is each man's quest for authenticity. The only way a person can find truth, happiness, and peace is by individuating their authentic self in their temporal context, their historicity. To blindly embrace a prefabricated truth embodied in groupthink social dogma is a sin of sloth that leads us into the devil's snare of inauthenticity.

Dogmatic religions or ideologies, including Islamic and Christian orthodoxies, groupthink Marxist secularism, and totalitarianism, impede authenticity by preempting it. Individual truth becomes pointless where the quest for truth is achieved by the group, where Godhead is manifest once and for all time in "the leader." With Islam the leader who possesses the authoritative channel to Godhead is Mohammed, or it is Hitler for the Nazis or Stalin for the Soviets. For contemporary PC quasi-Marxists, it is current group obsessions whipped into fury by self-serving demagogues. The *cause du jure* is treated like eternal truth, until displaced by the next *cause du jure.* Believers must submit to the group truth as they merge into a collective ego centered on their leaders. The problem is as old as civilization. The pharaohs declared themselves gods as did the Roman emperors.

A fundamental difference between Christianity, especially Protestant Christianity, and the more ideologically oriented Islam and Marxism lies in their different assumptions on time and Godhead. While specific Islamic views of time may vary with eschatology, the Islamic deity's unqualified omnipotence makes time static and eternal. However, an element of flux slips in because absolute omnipotence means Allah inscrutable, and can be unpredictable. In the modern quasi-Marxist flux view of time, truth is whatever the herd is stampeding toward at the moment, or whatever the party leaders or Big Brother Media say it is. Regardless of historic aberrations and diversions into barbarism by Christianity's followers and leaders, within the basic Christian myth time is a process that creates truth and meaning by unfolding the archetypes of its Godhead. More than in other major religious mythos, in Christian myth Time and the universe are God's individuation. The West's supreme artists grasp this intuitively. It is among the West's artists, not its theologians and philosophers, where one finds the most compelling and authentic visions of its Godhead.

Myths tap the energies of archetypes. They are also metaphoric statements the cultural psyche makes about itself. Godhead, like the psyche, reveals itself through primary mythologems which become the great dreams of a culture, dreamt by germinal artists like Homer, Dante, Cervantes, Shakespeare, Milton, Dostoevsky, and James Joyce. Artists deepen and amplify mythologems and thereby fashion secondary myths. Whether myths are primitive or sophisticated secondary myths, whether they are religious, political, scientific, or artistic, we perceive and judge reality through the categories established by myths driven by active archetypes. The controlling ideas and values of a civilization are set forth in myths by and about their implicit, as well as explicit, manifestations and expressions of Godhead.

The conflict of civilizations that intensified at the beginning of the new millennium is a conflict of Godheads. In the rudimentary Islam promulgated by Jihadist terrorists, the Godhead seems patterned after an inflated, power hungry alpha male ego—or to be specific, the Jihadist's God at times resembles a deranged desert Sheik. That type of male ego is prone toward ethically problematic biases and aggressive traits, e.g. subordination of women, intolerance of dissent and divergent lifestyles, repression of other religions, refusal to recognize that truth is a temporal quest. These have crippled contemporary Islamic culture. According to its harsher critics, like Winston Churchill, they condemn it to perpetual decadence. The Old Testament deity Jung depicted in *Answer to Job* is similar to Allah: the Jihadists' Allah is that deity taken to a pathological extreme. However, in Judaism and Christianity there was no Koran or Mohammed to forever petrify the Truth. Instead of being written by one man at one time, the Old and New Testaments were written by many men under disparate conditions across a millennium. The great diversity of this grand chorus of voices encouraged Judeo-Christianity to tolerate freedom and individuality and created space for respect for other religions. Most important, it encouraged individuals to seek religious truth and helped launch western civilization itself on its quest for scientific truth.

The truth unfolds, that is one of the transcendent insights from the life of Jesus, but the seed of the idea is implicit in Judaic culture and its messianic myths. Christianity brought a new dispensation, which inaugurated a new tardis or era in Western culture. Conceptions of Godhead continued to unfold beyond the divergent emphases of St. Paul and the Apostle John, through major thinkers like Augustine, Aquinas, John Milton, Immanuel Kant, Cardinal Newman, and Soren Kierkegaard. They then moved outside the confines of orthodoxy with Friedrich Nietzsche, and equally important but less well recognized, Carl Jung and A. N. Whitehead. Nietzsche, Whitehead, and Jung opened philosophical and theological doors for Christianity and western culture to deepen its understanding of the archetypes of the Godhead which form the mythological and metaphysical basis of Western values. These innovative geniuses raised/created the possibility of another dispensation that could move us beyond the current decadence of our "anything goes, nothing matters" quasi-Marxist culture where shallow, crowd generated ideologies form an ersatz Godhead that satisfies none and debases all.

MEDITATION XVII: JUNG'S CRITIQUE OF THE JUDEO - CHRISTIAN GODHEAD

1. As we have seen, Jung uses "Godhead" to refer to the totality of the four-archetype deity (the Quaternity) he sees individuating (or coalescing) in Judeo-Christian myth and Western civilization. Godhead is the archetypal reality, as distinct from the canonical *persona*, of the deity that has dominated the west since Constantine. This deity encompasses the Father, the Son, and the Holy Spirit along with a repressed Fourth Figure whom Jung in different phases of his thought speculated must be either Lucifer or Mary. Since the shadow, or the psychic energies ego suppresses, and the feminine or the *anima* are absent from the Trinitarian deity yet are necessary for wholeness, Jung posited a fourth figure in the Godhead. While Jung's four-part Godhead mirrors his four-part structure of the psyche, its activity follows the three stage dialectic of individuation that mirrors the three basic temporal tenses: 1) the largely unconscious Father (parent) who manifests the rude beginnings of individuation and consciousness, 2) the polarized two hostile brothers archetype, Christ and Lucifer, and 3) the consummation in a synthesis represented by the Paraclete, the *anima* (or *animus*, Cordelia or Prospero) of God, who epitomizes God as wiseperson and muse.

Jung strove to move toward divine completeness and balance, and away from ideology and dogma with his Quaternal Deity. He believed that Quaternity better represents the dialectical process of compensatory challenge, dialogue, and transcendence of opposites that characterizes divine and human individuation. The orthodox doctrine of Trinity evades the mythological truth of an unfolding Quaternal Deity. Jung thought he discovered this truth in his studies of alchemical and hermetic esoteric works. His prophetic manifesto *Answer to Job* attempts to explain as well as proclaim his discoveries. Jung sought in Quaternity a truly uniplural *imago dei* that incorporates the psychological, philosophical, and theological advantages of both monism and pluralism, (monotheism and polytheism) in a process that includes each but is neither exclusively.

In recent centuries the hostile brothers archetype has imposed its dueling dualism across the spectrum of Western cultural and intellectual history. In the philosophical arena the dualism has energized two opposing metaphysical outlooks whose contention has shaped Western thought. Reduced to elementary formulas these outlooks hold: 1) The universe is fundamentally static, monistic, rational and perfect, a position associated with the Son, who embodies the logos, and with Christian orthodoxy; 2) the universe is flux, pluralistic, irrational and imperfect, a position linked to Satan-Lucifer-Prometheus and increasingly dominant in the modern world due to the rise of positivistic and Marxist materialism.

The Paraclete personifies the neglected soul of Western civilization and may provide the key to its transformation in the next stage of cultural individuation. Jung, distracted by the dogma of the Assumption of Mary and his desire to reach out to Catholics, failed to develop the implications of the feminine background and nature of the Paraclete. As mid to late twentieth century scholarship has shown, early Christians identified the Paraclete with Sophia and other feminine aspects of deity found in scripture and the apocrypha. Misogynistic Hellenizers rejected the early Christians identification of Sophia with the Paraclete on the grounds that God must admit no imperfection and therefore must have no feminine aspect.

Instead of recognizing Sophia as the original feminine nature of the Paraclete, Jung tried to introduce the feminine in the form of the mother archetype represented by Mary. This is not only mistaken historically, but it is mistaken psychologically and mythically. The feminine in the dialectic of Godhead must be what Jung identified as "the wise woman," a version of the *anima* archetype, not the mother archetype. Erich Neumann's magisterial work *The Great Mother* delineates the mother archetype in all its magnitude and, therein, demonstrates why the father and the mother are the initial stage of individuation that precedes the brothers. The wise woman is the feminine of the wise man. Together they represent the third stage of individuation personified by a grandparent, a sage, or a priest, or priestess. Shakespeare readily grasped the archetypal reality with his artist's intuition.

The Paraclete as anima, the feminine function, contrasts against the Father and the brothers, Christ and Lucifer, to form the yin and yang of the divine androgyne and symbol of self that is the Western Godhead. Here are the stages of the archetypes of the individuating self and Western

Godhead shown schematically with their links to Shakespeare's characters:

Stage 3 Wiseman -Wisewoman archetype. Paraclete. Anima (Animus) mediated individuated self. Cordelia or Prospero.

Stage 2 Warring brothers archetype. Son (logos) versus Lucifer-Prometheus; ego X shadow; Job X Satan; Lear-Job X Lear-Prometheus's final defiance.

Stage 1 Unconscious primordial parent archetype; Great Mother and Yahweh. Yahweh X Job; Lear's initial folly with his daughters.

The archetypes of the Whole Quaternity diagramed:

PARACLETE
(female)

LUCIFER - PROMETHEUS ---------THE SON-CHRIST
(both male)

FATHER-MOTHER
(primordial parent figures, male & female)

The diagram of the archetypes of the Holy Trinity reveals its imbalance and incompleteness:

FATHER (male)

?? -----------------|-----------SON (male)

HOLY GHOST (neuter)

Jung's hesitancy in identifying Lucifer or Satan as the dark shadow brother is understandable because Satan is a purely destructive figure in canonical tradition. But an archetype is a nexus of psychic energy, not a canonical dogma. An archetype has a wide range of resonances and manifests itself culturally in multiple forms and shades of meaning. It is never identical with any single image or manifestation. For example, the second brother archetype could appear as Prometheus, who, through defiance of the father (Zeus), becomes a creative, revolutionary force and benefactor of mankind by eliminating what is outmoded and obstructive. Prometheus, the fire bringer, represents creative destruction clearing the way for progress. Or the second brother could appear as a wholly destructive force, like Satan, or in other forms, such as the trickster.

Regardless of the mythic figures selected to represent the four archetypes of the Godhead, what matters most is the dynamism behind the dialectic of the Godhead. (In respect to dynamism and dialectic, Prometheus seems the better choice, though the power of Satan in the popular imagination cannot be ignored.) Jung suspected that the western Godhead's archetypal dynamism and Its efficacy in reflecting the individuating self explains Its psychological appeal and tremendous historical impact. *Answer to Job* attempts to detail the evolution of the Judeo-Christian Godhead in the western psyche. Jung rejected theological Godheads exiled off somewhere in eternity, or Godheads formed by abstract philosophical theories divorced from vital archetypes. Indeed, all civilizational Godheads unfold historically for Jung. Moreover, for Whitehead they represent Time as a living, creative force or the *creative continua*.

Like Moses, Jung led his followers to the bourn of the promised land, but God had his reasons for not granting either one entrance. In mapping the basic structure of the psyche, with his four primary archetypes Jung had schematized the Christian Godhead itself. Because he was reluctant to recognize fully that the Holy Spirit or Paraclete represented the wise person *anima* or the feminine aspect of the divine personality, he failed to grasp the full implications for Godhead of his own schema of the psyche. Had Jung read *King Lear* as carefully as his mentor Freud read *Hamle*t, Shakespeare might have shown him how to connect the divine dots. Indeed, neglect of Shakespeare may be one of Jung's more significant short comings.

2. Despite its limitations, Jung's critique of Godhead is heuristic and opens doors to wider insights. Here, derived from Jung, is the first principle for understanding the relationship of God to man: *God, the self of the universe, interfaces with man through each person's self.* We can know, even if we fail to recognize intellectually, that God is real in the sense that the universe has an ordering, coalescing force, time, that gives it structure, stability, value, and relational meaning. Likewise, the psyche, as Jung has shown us, has an ordering, coalescing force, the personal self, that mirrors God in giving structure, value, and meaning to our own lives. These coalescing forces allow for the creation of consciousness and

feeling which in turn can apprehend truth and create beauty. Feeling, meaning, value, truth, and beauty exist only in the medium of time and are effects of the homeostatic self operating in time. They would not exist in a static world because they require consciousness and consciousness requires time in order to be able to make the distinctions that keep everything from falling together in an undifferentiated blur of the eternal present.

Does God exist is the second most basic philosophical question; the most basic question is: do I exist? Descartes answered it: I think therefore I am. A corollary is: I feel, I sense meaning and discern value, therefore God, the ground and creator of meaning and value, exists. Problems arise when people want Actual God to be more than the possibility of value and meaning in their world, of significance in human life, and of individual's power through free will to shape their lives and make a difference in the world. They want to give God an ego like their own, only more powerful. They want to see God as a big brother, or a parent, or a mentor or pal who especially favors themselves. They want to be like Joseph, the favorite son who gets the cloak of many colors. Trying to make God bigger, they diminish Him. Giving God inflated human roles is the source of endless misdirection in religion and among the religious. God is not the *totaliter alter*, yet God is the *totaliter maior*, the totally greater.

We fashion tribal religions that elevate some values and forms of meaning and deny others with the implicit objectives of elevating the tribe members and tribal groupthink, and setting them apart from and above outsiders. The tribal mentality seeks a god that gives status to the ego by offering it power, rather than a god that gives meaning to the self by offering the possibility of authentic individuation. The ego is attached to the tribe or group and seeks to identify with its power. The self is founded on the larger reality of Time as God. It seeks authentic individuation through a broadened and deepened experience of Time and Actual God in the context of our lives and their historicity.

MEDITATION XVIII: CHRISTIANITY'S QUEST FOR TRUTH AND LOVE

Some say God is like the Sun and the religions of man are divine sunlight refracted through a cultural prism, each religion having part of the spectrum, none capturing the whole. It's a useful argument for ecumenism and universal toleration, but that does not make it true. If all religions are equal, how do you explain their very different cultural, economic, political, psychological, and social effects. How do you evaluate their conflicting claims or choose among them? Studying them archetypally, you can see that they arise from different, though related, sets of archetypes and myths. The way they constellate and prioritize archetypes reflects their different historical and cultural contexts. Jung believed that, while it is valuable to study all religions, individuation and wholeness are best served by staying within the parameters of one's own culture or civilization. Going too far afield, losing historicity, can engender personality splits that impede individuation, and foster inauthenticity.

Having lived my three score and ten plus a few bonus years, traveled throughout much of the world, read widely, done differing work, known a broad array of people, and felt redeeming joys as well as endured my gamut of sufferings and injustices, my life experience leads me to conclude that the Judeo-Christian Godhead and its archetypes and mythical structures illumine mankind's most viable paths toward truth, wholeness, and ethical behavior. The dual emphasis on love and truth of the Gospel Jesus make credible that he and his life, as developed and expounded throughout the cultural canon of the of West, rather than by the churches in their doctrines, are the most universal manifestation of deity to mankind. Indisputably, they are the crucial manifestation of what has become the Western Godhead. The singular achievements and unique successes of Western civilization in science, philosophy, literature, the fine arts, technology, medicine, and in defending and expanding human rights, attest to the power of the basic myths and archetypes of the Godhead which guides our Western civilization and inspire its values. Yet Christians and Jews can learn from the other religions and remain true to their faiths.

Nonetheless, we must avoid the sentimental and intellectually sloppy belief that the Judeo-Christian, Islamic, Hindu, and Buddhist values are essentially the same. That blithe assertion is belied by their sharply divergent impacts on the morality and the intellectual and artistic activity in their host civilizations. For examples, contrast the rights of women and gays in the West versus Islam, or the treatment of animals and nature in the Hindu-Buddhist-Taoist East versus Islam and the West. Contrast the planet transforming scientific and technological dynamism of the West versus the stagnation elsewhere. While every great civilization produces outstanding art, that art is unique to its civilization. Each of the major religions appears to do some things better than or not as well as the others.

However, it is the stress on a Godhead defined by Love and Truth, over God as power in Islam, or God as Everything in Hinduism, or as blissful emptiness in Buddhism that appears to give dynamism to Western culture and motivates its achievements in ethics, thought, governance, technology, the arts, and sciences. Certainly, this crucial stress and its underlying insights have given the West major advantages. Christianity, and especially its founder's teachings and life, has provided compelling iconic representations of our core ethical values of truth and love. Crucially important, Jesus's life exemplified the heroism to pursue and the need and the courage to sacrifice for these values. By pursuing Judeo-Christian values Western peoples achieved signal progress in improving the physical conditions of human life, and they championed an end to slavery and freedom for all humanity. The idea of progress is close to being unique to the West. Freedom in all its richly varied expressions has flourished here to a degree that is unrivaled by other large civilizations.

(Jesus, Brazilian Sculptor, Aleijadinho, of Ouro Preto)

The founder of Christianity, Jesus, is the first of a long line of embodiements of one archetypal aspect of the Western Godhead. Many others have become Icons of the same aspect: e.g., St. Francis and a host of saints, and literary figures like Don Quixote, *The Brother's Karamazov*'s Alyosha, and *King Lear's* Cordelia. Cordelia is the supreme literary icon. She is not a realistic character, but rather a symbol of the union and interdependence of Truth and Love. As I have said, Cordelia as Truth represents the Paraclete in the Godhead which is the summation of Its revelation. Her sacrifice and her love also make her the Christ figure in *King Lear*. Cordelia, the paradigm of wholeness, can look squarely at any reality, however terrible, and maintain dignity, patience, and serenity without support from belief in or hopes for compensations in an afterlife or other divine interventions to ease her burdens or save her skin. In his re-union with Cordelia, Lear admits that he has given her cause to wish him harm. She forgives him without qualification with the quiet "no cause, no cause." Yet Cordelia never expresses regrets for speaking the truth or claims any responsibility for Lear's fall. For Cordelia, as symbol of truth balanced and harmonized with love, these virtues are the same because the highest love requires the purest truth. And so the West's supreme literary work dramatizes Christianity's most profound insights.

All religions hold some truth, but not the same truth or equal value as guides to all areas of human life, nor are they equally helpful or relevant for every culture and individual, or for all times. The most persuasive case for the truth and value of Christianity is not to be found in any theological tract but in the works of Christian poets and artists, especially the foremost one. In *King Lear* Shakespeare offers the most powerful dramatization of Christianity's Godhead and its fundamental values of love and truth. Shakespeare vision of Godhead is not a Godhead beyond time. Like Lear himself, his Godhead is wracked on the fiery wheel of time. It is represented in its myriad aspects by the greatest artists, thinkers, and heroes of Western history and culture. It began to unfold in the Arabian desert with Job, in Greek myth with its sublime rebel Prometheus, and reached a climax on a cross on Golgotha. Many centuries later, the paramount human artist created its most complete revelation so far.

MEDITATION XIX: GODHEAD AND THE EMERGENCE OF WESTERN SCIENCE AND TECHNOLOGY

In the Hellenistic period science and technology appeared to be moving toward the threshold of a breakthrough they did not again reach until the Renaissance. Archimedes, Aristarchus, Hipparchus, and Pliny were true scientists the like of which would not be seen again until Tycho Brahe, Copernicus, Leonardo, and Galileo. Historians struggle to explain why the promise of the Hellenistic era faded and then collapsed. Frequently, they blame social conditions such as reliance on slave labor and rule by despotic governments. These doubtless contributed to dampening the scientific spark. Another factor may have been the rise of fatalistic stasis philosophies and religions which exiled God to eternity, thereby devaluing the temporal world and this life. Fatalism precipitated a culture wide despair that spurred a turning inward in search of a soul, along with lowered interest in the physical world and how it works. People longed for order and the stable cultural values that seemed to be fleeing the world of the dying Roman Empire. Islamic expansion, plague, disruption of trade, and feudalism deepened the despair and accelerated the turning inward.

Yet the West shifted course in the middle ages. They were marked by slow but steady and significant technological advance that prepared the way for a turning outward toward the more adventurous and experimental approaches in the Renaissance. These changes were catalysts to an array of spectacular cultural missions, crucial among them modern science and technology, and the magnificent ideals/goals of progress and freedom. The West's peculiar view of Time, as real and irreversible, its time bound Godhead, and the high value put on this earthly life spurred development of a unique tragic a vision. These factors help explain why the West turned outward to develop science and technology. Moreover, they may offer useful clues to the historical options that await us.

Disparate major civilizations, the Egyptian, Chinese, Mayan, and Incan, independently developed their own sophisticated technologies. The technology of the new world groups, isolated from each other and entirely separate from the Afro-Eurasian landmass, evidence an inborn human capacity for technology. The human mind and our multipurpose hands facilitate the use of and direct us toward technology. Science is different, it is more the special gift of Greek culture with its Socratic quest for truth and for evidence of order in the universe. We might say that Socrates is the father of science, except acute scientific minds were already present in Hellenic culture. Socrates being the patron saint and avatar of philosophy and its quest for truth also became the guardian angel of science. Yet Hellenistic culture failed to sustain scientific endeavor.

Why did the Christian culture emerging from the middle ages into the Renaissance eventually succeed where the Greeks, along with the

Romans and many others, failed? Other ancient cultures had astrology and alchemy; only in the West did these develop into astronomy and chemistry—why? A major part of the explanation lies in the Christian Godhead. Five mutually reinforcing factors nurtured a cultural environment unusually favorable to the development of science: 1) qualified monotheism or unipluralism, 2) dynamism manifest in a drive toward progress, 3) the rationality of God and His universe, 4) the idea that man's free pursuit of truth can provide a path to salvation 5) a vision of a God in time rather than in eternity. The opposition of Christian society to slavery, stemming from the commitment to individual freedom and dignity, was a crucial sixth force that spurred Christians to look for ways to reduce human labor and turn mind and soul deadening tasks over to lifeless machines. Opposition to slavery is one of Christianity's supreme gifts to mankind, a gift that promoted human dignity and encouraged as well as helped to define personal and political freedom.

1.) QUALIFIED MONOTHEISM made the universe one, under one god with one set of natural laws, instead of a chaos of many gods pulling in different directions. The Judeo-Christian God is a developmental, process deity manifest in history and bound in time. The same applies to an even fuller extent to the Quaternal Godhead, the archetypal reality behind the theological *persona.* Christianity is qualified monotheism, or more accurately unipluralism, at a mythic if not a canonical level. The Judaic derived mythic-archetypal background, that makes the Christian God a process deity, controverts stasis based theological assumptions derived from Greek Neo-Platonists. These static monistic dogmas were at loggerheads with the dynamism inherent in

Judeo-Christian myth, which, as Jung and Whitehead grasped better than modern theologians, made the Godhead a developmental process that mirrored, on a historical platform, human individuation. Mythic dynamism first began to challenge theological dogmatism in the Renaissance and Reformation and moved to prevail in succeeding centuries. Absolute monotheism inhibits freedom and the free inquiry of science. A God who limits His power to allow for creativity across time leaves breathing space for the individual freedom, responsibility, and innovativeness that have facilitated the rise of science and technology.

The Renaissance and the Reformation, along with alchemy and Hermeticism, inaugurated a revival of more ancient images and archetypes of Godhead. The revival sparked a drive to rediscover and rethink Godhead and everything else—including science and technology. A key ancient image and archetype inspiring these changes was Prometheus, the fire giver, who represented an aspect of the Greek Godhead that challenged Zeus, the father figure in their Godhead. Fire represents truth, and Prometheus is the truth bringer. As we have seen, Shakespeare gave a crucial role to the Promethean archetype in his revelatory drama of the Western Godhead, *King Lear*. Prometheus represented a tectonic shift from Satan, the destructive brother of the hostile brothers archetype, to a creative or revolutionary destruction that clears a path for new good. In subsequent ages the Promethean archetype became a powerful driver of both technology and political revolutions against the oppression of a stagnant status quo. Prometheus embodied the psychic dynamism that motivated men to master nature and seek freedom from all forms of slavery.

2.) DYNAMISM - Although the awesome monuments left by the Incas, Egyptians, Romans, Khmer, and Chinese bear witness to a common human drive to develop technology, theirs were static technologies. Something happened in the Renaissance that would transform human life and with it the planet: technology became dynamic, and it acquired a sustaining motivation and a moral goal--Progress. Inventors, like Da Vinci, became cultural icons, and innovation itself became a way of life, indeed a way of salvation for Western society. Why did innovation take off during the Renaissance-Reformation while progress in the middle ages, although real, was much slower? The reviving Promethean archetype, coupled with Gutenburg's making Judeo-Christian myth available to all readers, gave new dynamism to the Godhead. The Bible stressed how God acts in history. For Protestants especially, history becomes a story of Progress. The quest for salvation itself is seen as a progress, for example in John Bunyan's highly influential *Pilgrim's Progress*. As science and technology began to improve the lot of the common man in the eighteenth century, Progress became a widespread and credible secular and moral goal.

In modern liberalism, Progress displaces traditional religious strategies to become the path to secular salvation. That made liberalism a force for constructive change. Where liberalism went astray later was in following Marx to elevate the group over the individual, and displace personal responsibility and individuation with crowd psychology and herd conformity. The result was denigration and betrayal of individual freedom making men slaves in the mass or the herd "owned" by totalitarian bureaucracies that posed as protectors of the masses. Despite Marxist and other diversions, the goal and ideal of Progress remains a dynamic influence for improving the conditions of human life.

The pursuit of technological improvement and expanded human rights, that is Progress, became a striving for individuation of and within the culture itself. Progress as a form of salvation began to supplant St. Paul's antiquated notion of salvation as atonement for original sin. Instead of repairing the damage of Adam's transgressions in Eden, Westerners began to focus on the task of perfecting an incomplete work—man and human society. Though seldom acknowledged, a major effect of the Renaissance and Reformation was a gradual rejection of medieval stasis assumptions about history and the universe for process dynamism and Progress. Under the radar also was the implicit rejection and replacement of St. Paul's original sinner, Adam, by the Son of Man and his Hermetic counterpart the Anthropos, the first or the universal man. These changes paved the way for Darwin, Nietzsche, William James, Whitehead, Bergson, and Jung.

The peculiar dialectical dynamism of the western Godhead engendered the idea of progress through the *coincidentia oppositorum* implicit in the two hostile brothers. Christ and Satan (or Prometheus), the bifurcate archetype at the core of Godhead's second stage, powers the dialectical dynamism of the Godhead. The hostile brothers are a throwback to the dualism of Zoroaster who was also the first great exponent of the idea of progress. Prometheus, who stole fire from Zeus and gave it to man, is the great mythic champion of human progress and freedom. His revival in Renaissance consciousness transforms the rebellious brother in the Godhead from a devil figure into an essentially benevolent, creative force driving the advance of science and technology.

3.) DIVINE RATIONALITY and the concomitant search for truth was derived expressly from Greek philosophy. The Godhead owed rationality to the early Christian thinkers' acceptance of Greek philosophy's and with it Aristotle's conclusion that God must follow and cannot change the laws of nature and of the universe He created. These laws are God's laws. Once created, they became inalterable features of reality. Stable laws require that time must be irreversible, that the past cannot be changed. Logic also requires irreversible time: A must always equal A. Stable laws based on a stable, unchanging past are essential to the pursuit of scientific knowledge, and to reliable scientific prediction upon which western technology depends. While we take the laws and their predictions for granted in the West, they are an alien notion to pre-scientific cultures that lack Western conceptions of irreversible time and a rational deity. Many Islamic theologians, for example, believe irreversible time imposes an unacceptable limit on Allah who, being absolutely powerful, must be able to change anything at will, including the past as well as logic and natural laws.

4.) INDIVIDUAL FREEDOM made the free pursuit of truth a path to salvation that began to be recognized and followed in the Reformation. By exploring and investigating God's creation, Christian scientists, like Isaac Newton, believed that they were furthering understanding of God. That was a powerful motivation for pursuit of truth in general and scientific truth in particular. The message of the Christian

God has been epitomized in a simple soundbyte: God is Love. But a second soundbyte became equally critical: "Ye shall know the truth, and the truth shall make you free." Jesus's promise (the key is what "truth" meant to post Gutenberg Christian Bible readers, not what it meant in the original Greek or even whether Jesus actually said this) makes truth a path to and requirement for salvation.

The truth that makes you free developed into individual freedom to pursue self-realization along with technological, social, and scientific progress. Truth that frees you to realize your whole, authentic self is a more meaningful and more heuristic path to salvation than is legalistic atonement for mankind's collective "guilt" incurred by a blunder in Eden. St. Paul's path to salvation depends on the myth of the fall of man which is challenged by modern discoveries about geology and evolution. Contrariwise, Jesus's promise of freedom through truth, and his respect for the value of the individual knower of truth, offers the hope of saving us from ignorance and error and of opening a path to enlightened self-realization. Seek the truth, therein lies salvation and freedom—as well as a powerful spur to scientific and technological progress.

The idea of freedom as salvation gave unique value to freedom itself and it motivated individuals to try to change their lives, their society, and the world. Historians have shown that science and technology began to first advance and prosper in the cities in northern Italy and Flanders, led by Florence, Genoa, Antwerp and Bruges, where the citizens enjoyed a high degree of personal freedom. Progress halted when those cities fell under the sway of despots, and it never started where despotism prevailed.

From the late middle ages until today, progress in science and technology, as well as general prosperity, have thrived most where freedom thrives. That is one of the prime lessons of the long cold war with the Soviet empire which the West won through scientific and technological superiority capitalized on by two determined leaders, Margaret Thatcher and Ronald Reagan.

5. THE GOD IN TIME made freedom and dynamism possible. Neither can exist, let alone flower, when men believe that value resides in a static eternity, or that existence is a chaotic flux. The God in Time makes possible the tragic vision that gives value to this life and the hope for progress . Together these spur the development of science and technology.

So in the Western view, God, the ordering force of the universe, became a unified, rational, dynamic, and progressive deity who encouraged followers to seek truth and freedom in order to improve themselves and society. Time itself is also unified, rational, dynamic, and progressive, in this cultural context. We can know God and become in harmony with God by understanding and living in synch with time.

A seldom recognized consequence is the power of Western technology, sweeping the world today, to carry with it Western ideas about God. Here we have a profound reason for Islam's deep seated resistance to and distrust of the Judeo-Christian West. Implicit in the West's all dominant technological culture are alien ideas about the nature of the world, time, our lives, and above all God. Indeed, Western technology is the creation of a rival to Allah, a process deity that accommodates

Yahweh, Christ, Prometheus, and Sophia in its unacceptable uniplurality, dynamism, and rationality. Moreover, the ennoblement of individual human freedom is a direct challenge to Islam which means submission and demands, above all, unconditional obedience to Allah and the Koran. India, China, Japan, and the pagan East are more comfortable with accommodating the technology of a foreign god whose mythological template, or Godhead, is less monolithic, and more comprehensive than Allah's. They are accustomed to multiple gods and have historically absorbed, rather than resisted, new gods. It has been, therefore, easier for them to embrace and utilize western technology. We have yet to learn how fully they will grasp and how effectively they will pursue the ideals of personal and political freedom. India and Japan support that hope, China, Thailand, Burma and others so far do not.

MEDITATION XX:
THE SOUL IN TIME

For Jung the self and the soul are to the psyche as God is to the cosmos or to life on Earth. Both the self and the soul unify and give value to the whole person, the self being more a thinking-intuition unifier, the soul being a unifier through emotion and sensation and the creator of value. The terms self and soul are related and sometimes interchangeable. But there are differences, soul speaks mainly through *anima* and *animus* while self can speak through any archetype. Although to refer to a group soul is ordinary usage, it would be odd to refer to a group self. The self is closer to being like an ego, that is, an individualized strategist and regulator that often works by constellating opposites. The soul is more connected with others and its community and tends to be a bringer of harmony.

The soul is an image maker par excellence, it is the dynamism behind imagination. The ego is the actual decision maker, the captain of the ship, in the process of individuation and soul making. The self is the "ego whisperer;" it's larger, more comprehensive than the ego which is but a single function of the psyche. The inflated, or hubristic, ego attempts to usurp the homeostatic self, a common pitfall leading to nemesis or disaster. The self must coordinate the various functions of the psyche, e.g. ego, persona, shadow, and anima-animus. It constellates opposites to correct imbalances and restrain excesses. The self's prime goals are individuation, harmony, and wholeness. But what is the soul of the soul, its essential character? Confusion about the soul in Jung's writings reflects widespread confusions about this important, though amorphous concept itself.

In the West there are two major conflicting conceptions of the soul, one derives from platonic stasis philosophy, the other from Hebraic and early Christian mythology where the implicit philosophical stance resembles process. (Materialistic flux philosophy tends to deny or question the existence of a soul.) The two conceptions of the soul entail two different conceptions of time, stasis time grounded in eternity and time as process. There is no place for an unchanging, indestructible immortal soul in either the basic Judeo-Christian mythology or the scientific conceptions of time assumed by the theory of evolution and modern physics. Positing an immortal soul raises serious questions. Since the individuation of living things is the fruit of experiences and choices over the course of their lives, how can a soul that appears intact at conception be truly individual? If their unique personhood is not the self-created product of their experience and choices, how can they be free or held accountable?

The doctrine of an immortal soul has major political ramifications, like the belief that the soul enters the body at the moment of conception creating a person, or "a life." All abortion, then, becomes "the taking of a life" and those who oppose all abortion style themselves "pro-life." Here we need to ask what defines an actual person and when does a person emerge. Does a person's development begin in the womb and unfold across life, or is a person a platonic form that descends from heaven the instant a sperm fertilizes an ovum? Empirically speaking, a newly fertilized ovum is only a potential person, like an acorn is only a potential oak tree. Believing a biological life and a person are the same thing requires an act of faith. For Jung, as well as for Whitehead and Heidegger, development of personhood is an existential task involving free decisions that build character. Like most valuable things in our time bound universe, personhood results from bewilderingly complex development over an extended period of time.

Instead of asking serious questions, intellectually indolent or timid people blindly follow some convenient authority who recruits them to help oppose a rival authority. Commitments like these lack authenticity, they are all ego and/or shadow, but no self. Their belief lacks an authentic moral compass based, as it must be, upon character developed through decisions and life experiences where the potential for wrong choices, suffering, and defeat or destruction is always present.

Fortunately, common sense and ordinary language usage offer a realistic view of the soul that accords with process philosophy and the process view of time. When Billie Holiday sings "Body and Soul" she is not singing about the immortal soul of platonic philosophy and Catholic theology. Billie's soul is an imaging, feeling, and valuing function whose creations are far more richly nuanced and experiential than the sterile concepts of philosophers and rigid dogmas of theologians. Billie is referring to her entire character, the qualities and imagery of her personal experience, and her whole emotional perspective and spiritual life. Billie's soul is not a given from God, she created it out of her common humanity, her active imagination, and her choices, and it embodies her unique life with all its joys and suffering, suffering being the most potent element for Billie. God did not create Billie's soul, or W. B. Yeats soul, or Bach's, or

Shakespeare's, or Walt Whitman's, or El Greco's. Within the medium of time, they all created their own souls, along with Caligula, Hitler, Saddam Hussein. The body and soul Billie sings of and from is her personal body

and soul created in imaginative response to the challenges of her unique place in time.

We have soul music, but all music is really soul music—Bach, Beethoven, Mozart, Verdi, Wagner, and Coltrane, along with the great imaginative artists in every field, all sing of the soul and from the soul. All sing from their unique feeling essence, because feeling predominates in soul. Though feeling is not the entirety of soul, for soul involves the totality of our experience, our decisions, our character especially, and our very being. Soul encompasses feeling, thought, will, sensation, intuition, imagination, and the mystical sense or the *unio mystica.* Above all soul develops and matures over time, and sings of time, as Yeats put it in *Sailing to Byzantium*:

> Or set upon a golden bough to sing
> To lords and ladies of Byzantium
> Of what is past, or passing, or to come.

Animals have souls just as they have characters. Advanced animals, like apes and bears, have more developed souls and characters because they are more conscious. Primitive tribal people recognized that animals have souls and often worshipped them for that very reason. This is important: conscious beings have souls and soul creation requires and is a function of conscious decisions. A rock has no soul, nor does a virus because they have no consciousness and make no decisions. It is a sin to stifle, stunt or disrespect the development of any soul. Buddha, the Taoists, and the East understood this better than the Western monotheists with their God outside of nature and their prideful exaltation of man over nature.

Even though the soul is the bedrock of individuality, we speak metaphorically of the soul of a community, a nation, or a culture because these can form a coalesced consciousness analogous to a soul, or if you will a collective soul. Herding, flocking, and pack animals, like cattle, geese, and dogs display a primitive version of group soul where individual consciousness instinctively yields to and follows the group. Homo sapiens, being more conscious, are better able to exercise choice. We can go with the flow of the group, resist it, rebel against it, or encourage or even learn to modify and direct the group's energies. Crowd psychology is a development from animal herd and pack instincts. It is potentially very dangerous in the hands of evil leaders who use the power of language amplified by modern communications to stampede the human herd toward destructive ends. Marx-Lenin-Stalinism, Fascism, Nazism, Maoism, and contemporary quasi Marxist political correctness, which combines elements of the earlier totalitarian mentalities and focuses on race more than class, are historical case studies in malefic crowd psychology.

Being more conscious and less instinctive than herding or pack animals, human individuals have a choice of going with the crowd soul or doing their own soul building by making authentic, responsible decisions for themselves. Crowds seek power and egos are tempted to follow the crowd in its power fantasies, ignore the self, and abandon individual responsibility. The power of the crowd is the greatest intoxicant known to man; entire societies can become fatally addicted. Crowd psychology first represses and then displaces authentic individuality. Wars and other historic disasters result when crowd psychology escalates into mad power fantasies as with Hitler, Stalin, and Mao. The soul of the crowd may seem greater and more powerful than the individual soul, but that is an illusion of the power fantasy. The crowd soul is shallow, inauthentic, and reckless, the individual soul is deeper, richer, and it can be far wiser. We must learn to let our souls sing and beware of their voice getting lost in the uproar from the crowd.

Some believe the world has a soul or that living nature has a soul. A world soul would be different from an individual or group soul, yet conscious in ways that are difficult for us to comprehend. Evolution may be directed by a world soul, and perhaps evolution is creating a world soul. The world soul may be what some perceive as God, as with pantheism. The American transcendentalists, led by Emerson and Thoreau, held that divinity pervades nature. Emerson posited an oversoul that relates to all nature as our personal souls relate to our bodies. Plato, the Neo-Platonists, and Hindus held similar views, and James Lovelock became a leading exponent in our time. Jung's writings on nature support the idea that nature has a soul. We may be able to communicate or at least commune with the world soul in nature. Our desire to be in nature may be a desire to get in touch with the world soul and to gain spiritual nourishment and harmony from it. A world soul, like individual souls, would exist in a span of time and be self-created through some process of coalescence. Different worlds, like different persons, cultures, and civilizations, would have different world souls. In the hierarchical levels of deity, the world soul would pervade nature and its domain would be the entire Earth. It would stand above civilizational Godheads, but below the Actual God of the universe, The God in Time.

The individual soul grows in time, soul building requires time and distance. The soul is created in time by conscious decisions and responses to the challenges of its world. It is the coalesced sum of our feeling life, and the essence of our very existence within and across time. Suffering, work, genuine achievement, solitude, and joy enliven and enrich the soul.

Spirit, by contrast, is the active force, the energeia, associated with *animus* or the masculine element in soul. Spirit is the vital spark of life. Traditionally, it is thought to return to God the creator while the body returns to dust. Spirit sets the direction and speed of time, it powers the flow of time in sentient beings. Jung saw spirit as the non-material, energetic aspect of a sentient being and links it with consciousness, purpose and the *animus.* The spirit and soul are opposites of body. Paradoxically, they are interdependent with body forming a *coincidentia oppositorum* or yin and yang, if you will, of life. Spirit and soul on one hand and body on the other impact time differently. The time of body and matter is the time of clocks, the calendar, and nuclear decay; it is steady, predictable and uncreative. The time of spirit, and of soul is creative and unpredictable. It is time of dreams, visions; and synchronicities where intuition and feeling grasp connections not immediately apparent to rational thinking

MEDITATION XXI:
TIME AND EVIL

The term "evil" is ever surrounded with confusion. Indeed, evil is and causes confusion. Confusion notwithstanding, in post medieval Western culture there are four principle conceptions of evil: 1) evil as impurity; 2) evil as *privatio boni*, or deprivation of good; 3) evil as malice and envy or psychopathy, the will to debase and destroy others; 4) evil as a challenge to create new good. The first conception dates back to the beginnings of civilization, the second to St. Augustine. The third is exemplified in literature by Shakespeare's Iago, Milton's Satan, and Melville's Claggart, and it gained urgency in the dark history of the twentieth century. The fourth view is compatible with Jungian psychology and more so with Whitehead's process philosophy. It may offer our best hope for contending with evil.

These conceptions support four basic stances or responses toward evil.

1) Impurity must be avoided or destroyed, and every person who is tainted must undergo purification or be scapegoated and ostracized or

destroyed, the stance of those who stress will to purity and damn Satan and his followers. It is compatible with flux philosophy and leads to dualism.

2) One can minimize the reality of evil, the stance of Plotinus, stasis, and Augustine's doctrine of *privatio boni*, a stance that takes us out of the world of time and its challenges into a fantasy of eternal bliss, like a return to the womb.

3) Since the holocaust and the other horrors of twentieth century totalitarianism, the third view of radical evil, manifest in psychopathy and sociopathy, became a widely accepted conception and concern. In the middle ages psychopathy was believed to result from possession by the devil to be remedied by witch burnings or exorcism. Malice and envy, more common forms of psychopathy, were often cited in the Elizabethan era. Anxiety about radical human evil abated in the more optimistic eighteenth and nineteenth centuries only to re-emerge as urgent in the darkness of the twentieth century. The current stance toward psychopathy and sociopathy is to fear and guard against them like the psychic plagues they are. Stopping their spread requires identifying them and establishing a moral quarantine. Since the truly evil are adept at disguising their intents until their damage is done, detection and quarantine usually come too late to help their victims.

4) The fourth stance of Whitehead, and compatible with Jung, acknowledges that evil is real in the world. Jung maintained that evil is real in God; Whitehead thought that God limited his power to allow for free will making god responsible for evil but not blameworthy. A prime

challenge in either case is to transform evil into new good infused with consciousness. Thus, evil becomes not a reason to question God's charity but evidence for it: God's allowance (Whitehead) or participation (Jung) in evil creates opportunities to realize new and more conscious good. In this view, evil is analogous to gravity in three respects. 1) We are fated to struggle against it. 2) If it triumphs, as when gravity creates a black hole, it becomes destructive and results in irreparable losses. 3) Without the constant struggle against evil, we cannot develop moral muscle, or strength of character.

Evils that we can transform into new good, like the contents of the shadow, may be termed operative evils. Evils that we fail to transform because of deep alienation, love's opposite, become incorrigible evils, i.e. psychopathy and sociopathy. From a Jungian perspective, the only absolute evil is the uroboric counterforce to the self's drive to transform evil into new, more conscious good through individuation. Psychopathy is a uroboric evil. While the uroboric counterforce is an absolute evil in that there is no going back, no redemption from it, nonetheless it is not a metaphysical evil opposed to good, like the devil is in outright dualism. Jung, along with Whitehead and process philosophy, rejects metaphysical dualism.

The various stances toward evil have generated three basic ethical positions associated with a momentous array of correlations to important philosophical, religious, psychological, political, and ethical concepts and topics. (See my previous book, *The Unfolding God of Jung and Milton*, for a development of these ethical positions in the context of Jung and Milton.) The table schematizes the alignment of the basic positions:

	Ethic of Repression	**Ethic of Transformation**	**Ethic of Rebellion**
Philosophy	Stasis	Process	Flux
	Essentialist	Developmentalist	Existentialist-Materialist
	Deductive	Synthetic	Inductive-Reductive
	Circle	Spiral	Arrow
Time	Past	Present	Future
Key Archetypes	Son-Logos	Paraclete-Sophia	Lucifer-Prometheus
	Ego	Self	Shadow
	Cultural hero	Wiseman-Wisewoman	Rebel
Key Ideas	Monotheism	Uniplural Quaternity	Atheism/ Agnosticism
	Absolutism	Contextualism	Relativism
	Dependence on God	Interdependence with God	Autonomy
	Privatio Boni	Evil 1) as Challenge to create new Good, 2) Uroboric evil	Impurity
	Hierarchy, Chain of Being	Creative Evolution	Anarchy
	Plato's forms	Concrescence Coalescense	Entropy
	Summum Bonum	Coincidentia oppositorum	Warring Dualities
Values	Cultural Canon	Individuation	Ideology
	Repression	Transformation	Conformity
	Comedic	Epic-Tragic	Satiric
		Human Canon	

While Jung's psychology aligns with the ethic of transformation in regarding evil as a challenge to create new, more conscious good, in its case through transformative individuation, it also stresses confronting evil as psychopathy in the notion of uroboric evil. Indeed, the uroboros is crucial for Jung, and the mythological formulation of it seems peculiar to Jung. It appears to be a concept that is foreign to Whitehead and Bergson. However, uroboric evil, linked to alienation, but without the mythological trappings, does seem consistent with Heidegger. Uroboric evil is a heuristic conception of evil that so far Western thinkers, including Jungians, have largely neglected.

The uroboros provides a valuable concept for understanding time's irreversibility. Uroboros takes for its symbol a single snake eating its tail, which represents self-containment and self-sufficiency, in an attempt to escape irreversible time. This contrasts with the symbol of interdependence, two snakes eating each other's tail or the Tao, or Yin and Yang, or the *coincidentia oppositorum*. Jung, as we have seen, maintained that the self constellates opposites to launch a dialectic that advances individuation. Uroboros, the devourer, works against the self and consciousness by promoting inflation, or hubris, which leads to alienation and rejection of interdependence with the *unus mundus* and God. The devourer is an implacable enemy of the self and of consciousness, which involves acceptance of and creative interdependence with a power outside and greater than the self, *irreversible time*. The devourer is, therefore, an absolute evil. Still, it has a crucial function because the struggle against it is necessary to the growth of consciousness and to creation of new good which together help redeem and give meaning to the irreparable losses and suffering inherent in the processes of time.

In a temporal, process universe where life is possible along with consciousness, creativity, love, value, meaning, and free will, there will also be tragic loss, ignorance, suffering, injustice, death, and evil. It is the tragic nature of all things in time that nothing endures, all must change, and change is never guaranteed to be for the better. The only true constant is the process of time itself. Every joy wanes and perishes, hopes are disappointed. All creatures suffer pain and defeat. Injustice and evil rage across the world. It may be a world striving for progress, but the road to progress is marked by fleeting victories soon displaced by new and more difficult challenges. Theologians puzzle over how a perfectly good, all powerful, all knowing God could create/allow so much evil and suffering. The answer is that God is not perfect or necessarily eternal, and rather than being all knowing God is learning and unfolding in a creative temporal process that gives conscious agents freedom to develop authentic individuality. God resides not in eternity, but at the knife's edge of the creative present.

Sacrifice has always been a crucial feature of religious practice. St. Paul and Jung both demanded a sacrifice made by God. For Paul man's sin of disobedience required a ritual slaughter of God's Son. In Jung's interpretation of the myth, Yahweh's own sins of blindness required him to suffer through the life and crucifixion of Christ that He might advance His consciousness beyond that of the amoral tyrant who allowed Satan to torture his servant Job. Job of course represents everyman. God's true sacrifice, with Jung, is to renounce the false, inflated claim of being the *summum bonum* and live within irreversible time in order that he may understand existentially the evil, pain, and injustice, as well as enjoyment and meaning, that goes with his gift of freedom to man and to creation. Jung's God is like King Lear and Prospero combined in a single character. God-Yahweh suffers like Lear so that he may understand man's lot which prepares him to know truth, meaning, and freedom, like wise old Prospero.

Actual God is not alone in time, but would be alone in the eternity of the theologians. Did Actual God create time and the universe because He wanted, indeed needed, company to become real, to be fully alive? (Prospero, the veritable God of his desert island, used his magic to wreck a ship on his shore in order to get more company!) Did God want/need company because the company of other free agents alone could give him relationship, meaning, and joy, and perhaps most important give him opportunity to love? Seen this way, as Whitehead does, creation was an act of supreme love. Somewhat like Prospero, Whitehead's God freely chose to limit his own power and to suffer in the world by creating other free beings He could love, and who themselves could give and receive love.

It is not possible to conceive good in the absence of evil, joy in the absence of suffering, justice in the absence of injustice, freedom in the absence of constraint, love in the absence of hate or indifference, and truth in the absence of falsity. Without loss, there rebirth, reconciliation, or reunion. Without suffering, there is no healing or deliverance. Without wrongs, there will be no mercy, repentance, or forgiveness. Without evil, good cannot triumph. Life is dull, null, and meaningless without *enantiodromio*, the dialectical dance of opposites. Many important concepts and values can be understood only in relation to their opposites. Indeed, in time things turn into or create their opposites. Where there is good the possibilities of evil are ever near, where evil rules there is always the challenge to create new good. Risk and uncertainty are the existential realities of the life of man, of all God's creatures, and of God Itself.

Theologians magnify the problems of evil and irreparable loss when they try to force the round peg of the process conception of time, implicit in the developmental Godhead of Judeo-Christian myth, into the square hole of Greek rationalistic, stasis, Neo-Platonism. In Neo-Platonic based canonical theology, God represents static perfection outside of time and the universe yet somehow ruling and creating them. As we have seen, Jung labored to show, and Whitehead understood, that the theologians' explanation of evil as *privatio boni*, or a mere deprivation of good, gainsays what our experience with history, psychology, and personal tragedy tells us about the world God rules and created. These orthodox Christian theologians, rather like their Islamic counterparts, thought it more important to protect God's power and perfection than to assert his love or to assess honestly the human condition.

If God is the *summum bonum*, omniscient, and omnipotent how could, or why would, He create a universe so unlike His perfect Self, so full of suffering, waste, injustice, and rank imperfection? If God created everything and knew all that would happen from the beginning, man could be neither free nor responsible for what he is and what he becomes. For man to be free, the existentialists recognized, he must create himself, fashion his character, by decisions he alone makes. If God is unchanging, omnipotent, and omniscient, everything that happens must be the result of divine decisions that stem from and express His eternally perfect character. The intellectually honest cannot avoid the conclusion that man is not truly a free agent under such a god. In short, God, not man, must be responsible for what man is and does. But man can be a lot of bad things, he can be a Hitler or a Stalin. So how can God be perfectly good, all knowing and all powerful and either create or allow a Hitler or Stalin? He cannot, unless you switch your theological framework to a process universe where God limits his power in order give man freedom and so that He Himself can unfold across time. In a process universe Hitler's and Stalin's free decisions make them ultimately accountable for what they do and who they are.

An unpredictable future is essential if God is to have a loving nature and man is to have freedom and individual dignity. Likewise, Time's irreversibility, entailing a fixed, stable past, is essential for accountability and the pursuit of truth and science. The process universe allows God and other beings, such as man, to unfold in time to achieve consciousness, beauty, value, and meaning. If God unfolds as a process in time, then the good as consciousness, value, love, truth, meaning, and beauty is created within time. Good, thus, does not partake of eternity but, like time, good is ever fleeting, "the stuff that dreams are made of."

Time as process means time is irreversible and some losses can be irreparable. The same holds for time as flux, the difference is that there is no enduring value in flux because events do not partake of a meaningful process with a direction and goals. Irreversible time and irreparable loss make tragedy possible, indeed inevitable. The crux of tragedy, is ever the irreversibility of time and the fact of irreparable loss. Tragedy, particularly Shakespearean tragedy, reminds us that evil can win and the triumph of good is ever uncertain. It reminds us we live in a universe where our decisions can matter terribly. It reminds us we are responsible for who we are, even if not for what our lives become, because who we are depends on how we handle what life offers and thrusts upon us. It tells us that the irreversibility of time, not death or eternity, is the ultimate reality of the universe and our lives. But tragedy means we have a chance to create value for ourselves and meaning for our lives through authentic responses to the challenges of irreversible time. Tragedy means that our decisions determine who we are.

In a flux universe, like the one playwright Samuel Beckett depicts, evil is far easier to explain than it is in a stasis universe. Since all is ultimately chaos, chance, or worse anarchy, rules in flux. The coin does not always come up heads, things often go wrong unpredictably. As the saying goes, "sh-- happens." But where flux is deemed the ultimate truth, men do not find that truth emotionally satisfying: it seems a prescription for meaninglessness. In a process universe, however, evil becomes a challenge to create new good and therefore it has the potential to become meaningful. That new good appears in the form of an expansion of order, harmony, complexity, and consciousness and in the creation of value, beauty, and meaning. The pivotal ethical questions become not, how to avoid evil, but how to transform evil into an authentic new good and in so doing to foster personal individuation and the overall unfolding of consciousness.

History is replete with disasters and tragedies, both natural and human induced. Most societies have psychopathic members, though they vary in frequency as some societies and some timeframes are more fertile grounds for psychopathology than others. Cultures, political ideologies, and religions often entertain and encourage psychopathic beliefs, doctrines, practice, and leaders. Psychopathy can spread throughout an entire society, as under the reigns of Hitler, Stalin, and Mao. Or it can become a prevalent force in society, as with racism, anti-Semitism, homophobia, or terrorism. History shows us that there is no avoiding the challenges of evil. Evil in some forms is ever present. Like Hitler, evil can emerge from seemingly nowhere as an ominous force in our lives.

Der Führer, rare photo prior to his demise

Psychopaths offer windows upon the psychology of evil. These purulent individuals sometimes appear to be born with a powerful attraction to evil. More often, though, the evil, as with Shakespeare's Macbeths, begin as apparently normal men or women and become psychopaths by yielding to the temptation to do wrong knowingly in order to gain status, wealth, or power, or just to protect what they have. Like Macbeth, the evil tend to withdraw from their own humanity into a self-created, self-contained world full of sound and fury signifying nothing. Macbeth and other fallen Shakespearean characters recognize their meaninglessness. For example, the dying Edmund in *King Lear* realizes his moral dilemma: "I pant for life. Some good I mean to do despite my own nature" (V.iii. 243-244). In *Hamlet* Claudius cannot pray or repent because he has become too attached to his evil course and its rewards. Not all the evil, perhaps not any of the truly evil including Hitler, share Edmund's dying impulse to do some good despite their natures. Certainly, Iago displays a chilling lack of regret:

> *Demand me nothing: what you know*
> *You know:*
> *From this time forth, I will never speak word.* (V. II. 303-305)

Like Iago and Milton's Satan, the evil withdraw from time as a creative process. Refusing to accept a world they cannot completely control, they retreat into their own fantasy world isolated from the judgment of others.

In his early poem, *Comus*, Milton captured the essence of this uroboric pattern of withdrawal from Time into the devourer's self-contained, self-fed, self-consumed prison of sterile psychopathology:

> *But evil on itself shall back recoyl,*
> *And mix no more with goodness, when at last*

Gather'd like scum, and setl'd to it self
It shall be in eternal restless change
Self-fed and self-consum'd. (593-597)

Milton's phrase "Eternal restless change" sums up uroboric evil and the flux sub-universe it creates: It is a moral version of a black hole. It also fits those who, attempting to escape the challenges of time in a process universe, seek refuge in the delusion of a perfectly self-sufficient eternal power only to find themselves trapped in an infernal flux of meaningless internal change.

The definition of evil, like the definition of God, varies with the underlying view of time. For stasis evil is ultimately unreal, a mere deficiency of good, not a power in its own right. Psychopaths in everyday society and the forces of psychopathology in history, along with disease and natural disasters, belie that rosy view. For flux, evil, and meaninglessness are just part of everyday reality: by the laws of chance things must often turn out badly and nothing is ever sure. Life creates chaos, impurity, and waste. But there is more to life than can be accounted for in flux philosophy. Life brings increased complexity, and increased consciousness as it struggles with time and its challenges. Evil and suffering become challenges to expand consciousness, find new meaning and value, and to adventure out into the realm of time and discover new purposes. Across time and history, evils fuel the drive venture forth in order to create new good, and thus they can further the unfolding of the Actual God of the universe whose creativity, goodness, and glory is interwoven within and throughout Time.

.

MEDITATION XXII: DIVINE ANACHRONISMS

Anachronistic tribal and canonical conceptions of deity maintain widespread, dystopic influence in world culture and politics. Notable examples are the temperamental, high maintenance Yahweh, the legalistic, honor obsessed deity of St. Paul who allowed his Son to suffer on a cross to atone for a minor infraction by two naïfs, and the absolute despot Allah bent upon forcing all humanity into the procrustean bed of a draconic Sharia legal code fashioned to restrain the barbarism of desert nomads. Increasingly popular in our consumer society are the sugar daddy or "happy Buddha" deities who offer shallow bromides and promise small favors in exchange for monetary contributions. These are unreliable, and counterproductive guides to modern life, its problems, and unique challenges and opportunities. They leave the soul hungry and the mind in restless discontent.

Responding to their influence, many westerners dismiss God as an outdated contrivance. They have lost sight of their inner God-image, a distinct psychological force manifest through a mythic veil, but with a metaphysical foundation in the nature of time. Frustrated with anachronistic tribal and canonical God-images, they renounce religion altogether and turn against their own Judeo-Christian culture. We just don't need it, they argue, ignoring the universal human need for connection and wholeness that religion and God serve. But what is to replace religion? —banal, politically correct dogmas of the left or the right? (The right has its own equivalent of PC, MC for morally correct.) These are expressions of the widespread frustrations with modern spiritual vacuity. The intractable truth remains that men need and seek religion in all cultures, even our relentlessly secular culture: to do so is ingrained in our nature by the conditions of human existence. As Jung learned from his patients, the god image is hard wired into the human psyche. We are conditioned to seek God because life forces us to contend with time irreversible and its irreparable losses, sufferings, injustices and fleeting joys. At the same time life affords glimpses of a greater reality that inspires wonder and awe.

Too often, in sloth, cowardice, and despair, we yield to the temptation to lose ourselves in the crowds following the popular groupthink ideologies. As G.K Chesterton observed, when people stop believing in God they don't believe in nothing, they start believing in anything. Still, in quiet moments we marvel at the amazing world around us. Why are we here? Why is it here? What brought it all into being? Does it have any purpose; can our lives have meaning, or can we create value, or find enduring truths? We gaze at the Milky Way, Iguassu falls, Denali, or look up inside the Sagreda Familia, or spot a beautiful animal in the wild, and we are filled with awe and long for higher meaning.

We wonder what created this world so full of majesty, beauty, and tragedy. We ask, who is man, what is he doing here, and what is a good life and how to live it? We fear sickness, injustice, loss, shame, emptiness, and especially death. We ponder what may lie in that "undiscovered country from whose bourn no traveler returns." What meaning does death hold for our lives, and for who we are right now? The East believed in reincarnation, that our souls are recycled again and again into this veil of tears. The West believed in an afterlife and/or a resurrection into paradise, but these convictions are waning. Increasing numbers of modern people are agnostics or outright atheists who assume there is little that we can be sure of this side of the grave, and nothing at all beyond the grave.

The youngest major religion, Islam, arose in the seventh century AD. The others are centuries older. All formed in times when the conditions of everyday life were radically different from what they are today in modern, technologically advanced societies. The average lifespan has tripled, the childbirth death rate for women giving birth and for children has plummeted. The human population has exploded from 200M in the time of Mohammed to 1.0B when George Washington lived to 7.2B today. Nearly all peoples once worked the land in primitive rural settings, whereas now we are crowded together in giant, impenetrably complex, heavily polluted metropolises. Yet many fearful diseases are now history. Plentiful, safe, healthy food and clean water, if not clear air, are readily available in advanced societies. Free "education," disregarding quality, is widely accessible. Levels of wealth and comfort for the affluent professional classes are extraordinary compared to those of the aristocracy of just 200 years ago. One can travel to any city on the planet in 24 hours, and communicate anywhere instantaneously. We enjoy the internet, television, time saving appliances, jet planes, modern medicine, and of course our ubiquitous smart phones. We look forward to space travel and colonization, to genetic engineering, life extension, further disease eradication. So why the anger and despair over this brave new world that Western culture, science, and technology have created?

Our advances would be inconceivable to Moses, Mohammad, St. Paul, Confucius, Plato and other seers whose unchanged ideas, derived from anachronistic life conditions and experiences, shape the religious outlooks that prevail today. The prodigious scientific discoveries of the last 500 years from Copernicus, through Galileo, Newton, Darwin, to Einstein, Hubble, and Turing were not around to inform the conceptual frameworks of the ancient seers who founded the major religions. All were ignorant of nuclear physics, evolution, organic chemistry, anthropology, cybernetics, artificial intelligence, and modern astronomy. None had any premonition of the scientific and technological progress that would radically alter human life and the planet we live on. What they did have was a firm conviction that human society and the conditions of human life they knew, the technology and science that was handed on to them, including man's ideas about the earth and its place in the cosmos, were permanent realities/truths that would not be subject to significant, let alone revolutionary, future changes.

Much of the seers' teaching is based on assumptions we know have limited relevance today or are just plain inapposite. Yet many oblivious modern people, still try to follow to the letter these oblivious ancients, simple naïfs compared to themselves. But worse and lesser than the seers, who often responded wisely to the challenges of life as they knew it, their contemporary proponents are too often uncritical ideologues. Today's fundamentalist followers of the ancient seers rely on their anachronisms to explain the nature of the universe, to set inflexible codes of moral conduct and social behavior, and to teach us the meanings of life and the nature of the human condition. No wonder so many of the disillusioned react against their own cultural heritages with frustration, anger, and despair, and turn instead to facile, high-risk political fads.

Anachronisms notwithstanding, some of the ancients offer invaluable guidance from their insights into human nature itself, which changes little. Jesus and Hillel, who taught us basic equity, gave us a reliable foundation for ethics that for almost two millennia the West accepted, at least in theory. Buddha's ethic has different mythical and metaphysical foundations, but it is similarly sound. However, Marx and acolytes, in a throwback to tribalism prompted by political opportunism, chose to view human individuals as group members, rather than as unique souls. Neo-Marxists replaced the Judeo-Christian and Buddhist ethics with group think, group guilt, and group power which could be manipulated to prepare the way for totalitarianism of both left and right. Marxism has generated much of the political chaos, spiritual malaise, and intellectual banality of modern times. Those who minimize the influence of ideas on history would do well to contemplate the immense and often deleterious impact of Marxism in the last century.

Jesus's corollary to "Do unto others," the elevation of truth proclaimed in "Ye shall know the truth and the truth shall make you free," acquired ever enlarging meaning across the centuries as it motivated the intellectually inquisitive to study nature and pursue technology in order to improve the physical conditions of human life. From this search for truth, scientific endeavor and the quest for progress evolved.

Buddha's call for universal compassion goes beyond "do unto others" which appears to limit ethics to humanity, implicitly downplaying the welfare of animals and the rest of nature. Buddha saw that "others" should include all living, feeling things, although that "Buddhist" extension is represented in Christianity, most famously by St. Francis. The Buddha's teachings, as well as those of Taoism, offer nature better protection against the ravages of humanity than is generally afforded by the monotheistic religions. Nonetheless, Buddhism and the east offer no equivalent of individual freedom, progress, or the search for truth that makes men free.

The evolving Western Judeo-Christian tradition stressed the value of the individual, and encouraged belief that intentionality and consciousness are unique to humans and their gods. However, intentionality and consciousness appear to be general principles informing the evolution of life, and possibly our time bound universe itself. This perspective encourages the hope that there may be a drive toward order, value, and meaning, inherent in the nature of time itself. While it may not accord with St. Paul and Christian and Jewish Orthodoxy, it accords philosophically with the founder's stress on love and truth and mythically with the active archetypes of the Christian Godhead. God is Love is explicit in the New Testament. God is truth may be only implicit in the teachings of Jesus, but it became a fundamental Christian value thanks to Greek philosophy.

Studying the seers and even revering them cautiously and selectively makes sense: we stand on their shoulders spiritually. Our challenge, however, is to learn from their universality without being misled by their parochialism. We must resist the temptation to turn their anachronisms into inalterable dogmas. Such restraint and circumspection requires placing them in the temporal framework of the unfolding of a universal wisdom greater than the personal wisdom of any one seer could ever be.

We need also to add to the seers' numbers. The Shakespeare opus offers a more reliable and comprehensive guide to understanding human nature and the human condition than any sacred text from any culture. Or perhaps one should say that Shakespeare's opus is, or should be, the most sacred of our sacred texts. But there are many other magnificent works in the Western canon, and the broader human canon. The wealth of mankind's entire canon is prodigious. The Bible asks "where is wisdom to be found?" The human canon is the "open Sesame" to Wisdom for those who approach it with diligence, humility, perspicacity, and without bias. We must endeavor to see through the eyes of all, listen to the great chorus of humanity singing across time itself.

Jung may be the most useful and universal modern Western guide to questing souls. He taught that man's only reliable path to divinity is individuation or the quest for consciousness. When authentic gods (or demigods) appear they are mediated from the self and serve the purposes of self, not the appetites of the ego. Beware of Gods that resemble patriarchal desert sheiks or power hungry Caesars, or greed driven self-help gurus: these are but egos deified. God and the self, Jung taught, become manifest over the process of time, not in a single moment or era. It is possible for a psychologically healthy religion, based on self, not ego, to unfold in a culture as the self can unfold in individuation and the universe unfolds in time. Some speculate that ancient Zoroastrianism, early Buddhism, and early Christianity may have approached being such religions.

The greatest religious challenge of our age is to continue the individuation of our religions by advancing their understanding of Godhead, Nature (or the Goddess), and ultimately of The God in Time itself. That will require broadened perspectives on our divine anachronisms that honestly acknowledge the limitations imposed by their historical contexts and our own. We must resist the ego's temptation to project specific, dated human norms and social/cultural practices unto deity, or to merely go with society for convenience or expediency. In short, we must move beyond the egolike gods of the tribe, the nation, and the sect.

"God" may have made man in his own image; if so, He likewise made the tiger, the crow, the scorpion, and the butterfly in some of His infinite array of marvelous images. Anthropomorphism, seeking literal analogies between the guiding force of the universe and human personality, is an emotional, theological, and ethical trap. Wisdom requires a Shakespearean metastance. Just as there is a destructive idolatry in sanctioning kings with divine right, popes with infallibility, and dictators with totalitarian powers, so it is idolatrous to worship either a past era's projections on deity or our own. This is a form of blasphemy that encourages the dogmatic intolerance behind much of the contemporary ideological madness. Whom the gods will destroy, they first make mad. Ideology is the dominant modern form of collective madness.

The mad hubris displayed in overpopulating our planet, and thereby degrading and endangering the natural environment that sustains us, is rooted in the idolatry of the anthropomorphic religions manifest in their deifying man over nature and raising their own parochial teachings above the human canon itself. Condemning rational limits on our population increase, callously denying a right to life to other species, and denigrating and disrespecting the natural order in the names of their deities, our anachronistic religions' hubris may finally drive them into well-deserved disrepute. But the price of learning could be an apocalypse in nature itself that would bring incalculable suffering to much of humanity. We may not need to fear the wrath of Actual God, the wrath of Nature is a different matter. Humanity is guilty of collective hubris in willfully disregarding the implications that the last five centuries of scientific and cultural advance hold for how we relate to the rest of nature. The anthropomorphic gods may claim to be merciful, but nature and the harsh limits of our time bound existence are never merciful to the blind, the hubristic, and the stupid. The penalty for hubris is ever nemesis

.

MEDITATION XXIII: CONTEMPORARY CHALLENGES

Whether we think of ourselves as atheists, agnostics, or devout theists, the basis of our beliefs about ethics and conduct remains rooted in our cultural backgrounds. The gods reign in our cultures whether we as individuals believe in them or not. To make choices that are compassionate, rational, and wise, we must understand the religious mythology underlying our core beliefs and values. Socrates said the unexamined life is not worth living. Likewise, the unexamined God is not worth worshipping, and the unexamined religion is dangerous to follow. Yet no time since Martin Luther nailed his 95 theses on the door of Wittenberg Castle in 1517 has there been so little energy devoted to understanding the grounds of accepted beliefs as today. Blind, slothful adherence to dysfunctional religious and ideological tenets, sanctioned by anachronistic prophets, inhibits moderns from adapting effectively and expeditiously to the existential and ethical challenges of our rapidly changing times. Here are eight salient examples.

1. DESECRATION OF NATURE. Theologies that divorce man and God from nature, and place ultimate values in eternity rather than in

temporality, have set modern man at odds with his natural environment, and condoned a genocidal cold war against non-human species. We hear an uproar about the dangers, real and imagined, of climate change. It misses the underlying problem which is not climate change so much as environmental degradation and desecration of nature. Climate change is only a problem if it causes degradation of the environment necessary for sustaining human civilization and for maintaining the diversity of biological life. Moreover, environmental degradation is a clear and indisputable fact. While the climate is always changing, the reasons for that change remain problematic. Mankind has only had a major impact on the environment in recent millennia. For millions of years before man, huge climate swings pushed the planet in and out of ice ages where the sea levels rose and fell by hundreds of feet.

We see environmental degradation everywhere as ever spreading human economic activity spurred by population growth consumes forests, pollutes the air, contaminates the seas and fresh waters, and destroys the habitats for rare and beautiful creatures and plants depleting the genetic treasury of earthly life, thereby despoiling one of The God in Time's masterworks. The deification of man above nature, making God in the image of man rather than in the image of life, is perhaps the ultimate blasphemy and paramount threat to human survival. Making God man is an act of hubris that may prove to be the nemesis of all earthly life. The major religions of today either endorse or fail to effectively discourage species selfishness, all the while deploring individual selfishness and demanding sacrifices on behalf of their institutions because they claim to serve mankind.

Humanity, a single species, is a small part of life on earth, and earth itself is an infinitesimally tiny part of Actual God's unimaginably vast creation. In the next level of religious and ethical development mankind must learn to live in harmony with and show respect for nature. Yet our reactionary religions preach species chauvinism which leads to the ruthless exploitation of nature and the callous extinction of other magnificent species. Human survival may depend on Yeats "second coming," a true revolution in religion that re-consecrates nature by extending humanity's ethical obligations to all life on earth.

2. PERSONS. Failure to recognize the temporal process essential to the creation of a human person has led to moral confusions and controversies such as those surrounding contraception, abortion, and stem cell research. These controversies form a fatal distraction from truly serious moral issues like our species selfishness and the resultant desecration of nature. Some zealots insist that the mere genetic plan for a human being in the chromosomes of a newly fertilized human egg must be treated as an actual human person. That is like making an architect's drawing for the Taj Mahal equivalent to the Taj Mahal. Morally muddled ideologues, following rigid dogmas, assert that every abortion must be banned to protect "human lives," meaning in most cases genetic blueprints. However, traditional ethics are not about protecting life in the scientific sense of DNA. They are about protecting persons, that is unique, existential individuals who develop through extended experience within their physical, social, and cultural environments, beings who can feel, suffer, love, hate, and think. Moses, Jesus, and Buddha knew nothing of DNA, but they knew a great deal about persons acting as responsible

members of society, and they understood the crucial roles of intention and consciousness in ethics.

Because the fetus begins to have complex experiences and feelings in the womb, it is not unreasonable to count as a person an infant in the latter stages of pregnancy, He/she has a nascent form of consciousness, feels, and has already, before birth, passed through the initial stages of personhood. It makes sense, legally and ethically, to protect fetuses in the latter stages of their development. But it makes no sense to give an undeveloped fertilized egg, devoid of thought and feeling, the legal and moral status of a person.

The philosophical background to the abortion dispute is crucial, but almost always ignored. Those who embrace the dogma that a person appears at conception believe an immortal soul attaches to the fetus at that point. Their belief reflects traditional metaphysical assumptions of stasis theologies. Those who find partial birth abortion morally acceptable are likely to assume a flux view of time and the universe. In the American constitutional system neither group has a right to impose their own theology specific ethics on the general public. The approach of process philosophy, stressing the unfolding of personhood, offers a solution that can avoid violating commonsense and the US Constitution.

Is God a person? Actual God is not a "person" in any ordinary sense of that word. The mystics have come closest to understanding this. God accommodates persons, but that is not the same as being a person. The differences between Actual God and a human person must be incomparably vast. An entity whose presence and sway stretches over billions of light years, must differ intrinsically from one whose life spans a few decades and whose sway is limited to a particular place and time on earth. Actual God is an *uberprinciple*, like gravity only over gravity, not an entity with an ego. Insofar as God operates like a human person, it is as a homeostatic function like a self, not a decision making, function like an ego. To attribute egolike predilections and choices to the Actual God of the universe is crude anthropomorphism. However, the civilizational Gods inevitably do resemble persons because they acquire the character of the civilization whose collective mind and soul they embody. Like persons they vary in degrees of individuation and the extent to which they are in touch with the self, in their case the nature deity of earthly life as well as Actual God.

3. LGBT SCAPEGOATS. In the name of religion, some assert that frozen human embryos are persons entitled to full rights, while adult LGBTs are, ostensibly, demi-persons with substandard or no rights. Denial of full human rights to LGBTs varies with the religion and its traditions. LGBTs have no rights under Sharia law, especially in Islamic countries where homosexuality is punishable by death. In practice LGBTs don't fare much better with orthodox Hindus and Jews or with fundamentalist Christians. Buddhism has a reputation for being the most tolerant of the major religions. Nonetheless, the Dalai Lama took his time to endorse gay marriage and condemn homophobia. Nothing in the core teachings of the Buddha sanctions denying compassion and full rights same sex lovers. Jesus was no different. While a passage in Paul's Epistle to the Romans

does appear to censure effeminate people, its meaning and implications rest on a problematic translation.

The golden rule requires us to love all humanity, even sinners. Same sex love is never discussed, let alone condemned as sin, in the four Gospels. What is condemned repeatedly and unequivocally is divorce. Yet contemporary Protestants mostly follow John Milton's "Doctrine and Discipline of Divorce" which allows divorce on the grounds of spousal incompatibility; and the Catholic Church has slowly tempered its opposition to divorce. While compromising on divorce, the Church and Evangelical Protestants at the same time hardened their opposition to homosexuality. Expediency may explain the shift. Living much longer today and having smaller families, spouses have more time for personality change and to grow tired of each other, and so they seek divorce more often. Sharp moral condemnation of divorce is a prescription for vacant pews and empty collection plates. The hostility to same sex love, where heterosexuals scapegoat gays for the guilt they secretly feel about their own sexual sins, like fornication, adultery, and divorce, is a textbook example of projection of the shadow. The very arguments Milton uses to justify divorce also justify tolerance of same sex love and marriage. Social attitudes change with the times, but "do unto others" is forever; and it proscribes for all time unjust bias and cruel discrimination. Shakespeare said it best; though he speaks of romantic love, his great sonnet applies to all forms of genuine love. The marriage of true minds is the veritable North Star of ethics:

Let me not to the marriage of true minds
Admit impediments. Love is not love
Which alters when it alteration finds,

Or bends with the remover to remove:
O no! it is an ever-fixed mark
That looks on tempests and is never shaken;
It is the star to every wandering bark, Sonnet 116

Bias against same sex love once had a purported rationale in society's need to maintain and desire to increase human population. I say purported because in pre-modern society homosexual and bisexual people usually married and reproduced anyway; many of those who did not became priests, nuns, or adventurers. Gays and lesbians living openly together and acting as a discrete social and political group is one of the most surprising and significant social precedents set in the 20th century. Yet today's LGBTs still reproduce or parent at a rate estimated to be almost half that of the general population. Disapproval of their being parents remains widespread, notwithstanding the lack of evidence that they are less successful at parenting than their heterosexual critics.

Focusing on the LGBT pseudo-problem is another distraction from truly serious problems. High among them is population growth beyond an optimal level for human health and welfare, and for maintaining the bio- diversity essential to stability in the biosphere that sustains all life. A disturbing, but largely ignored, effect of the resultant environmental degradation is the rise in general infertility affecting humans no less than animals. Conveniently ignored also are the ways the bias against LGBTs damages them, as evidenced by their much higher suicide rates, alienation, loneliness, and difficulties integrating into society. Is it not irrational to condemn sharply the damage racial bias inflicts upon African-Americans, Jews, and other minorities and yet assume that bias against LGBTs does no harm to them or society? We condemn human sacrifice, the taking of human lives as a religious practice. How can we condone the blighting of human lives as a religious practice?

4. RELATIONS OF THE SEXES. For the last several millenniums men have used religion to keep women subordinate. Subordination made the feminine influence in society and culture less visible, but it did not necessarily lessen it. Men must always contend with the women in themselves and in other men, the *anima*-- along with the man in their women, the *animus*. Most great poets and painters have been men, but much of their art has been inspired by women and even written by their inner woman, the *anima,* think of Dante, Petrarch, Henry James, and James Joyce, or Botticelli, Rubens, Renoir, and Picasso. Subordination of women and masculinization of the deity deprived the deity and its culture of wholeness, not to mention denying women, and men, the opportunities to develop and express their wholeness. Jung well understood the crucial importance for a culture's psychic health of recognizing the feminine aspects of Godhead. A culture that denies the feminine divine will be seriously imbalanced and produce unstable societies, often ruled by dangerously deranged men.

Economics, Marx taught us, finds ways to fundamentally alter entire cultures and civilizations. Ever increasing numbers of women entering the workforce and spending less time in child rearing over the last century and a half has created a cultural shift as profound as the shift to patriarchy thousands of years ago. We are entering an era where women have economic, social, and political power equal to men and, perhaps more critical, where the feminine principle can openly contest the influence of the masculine principle. Throughout the world all developed societies are marked by growing disharmony of the sexes as women and men struggle to find a new *modus vivendi*. Boys and young men face identity crises unsure of how to adapt to the changing roles and status of men in modern society. Young men are rapidly falling behind women at all educational levels internationally: the female to male ratio of new US college graduates is 57%/43%. The suicide rate among men is 3.5 times higher than the rate among women—what does that say about the challenges of men's lives in our societies? Leaders of the major religions have failed to provide guidance for those suffering crises around their gender roles. They have been reluctant to adapt to new understandings of human sexuality and their implications for religious doctrine and the civilizational Godheads. Fortunately, the Christian Godhead does not need to be re-invented. Christians do need to seek better understanding of its essential archetypes, especially the Paraclete. Here Jung blazed a trail that deserves deeper exploration.

5. GROUP PRIVILEGES VERSUS INDIVIDUAL RIGHTS. Exodus 20:5 *"I the Lord your God am a jealous God, visiting the iniquity of the fathers on the children to the third and the fourth generation of those who hate me."* Squaring the Old Testament Yahweh's vindictive ways with the New Testament Jesus's "Do unto others as you would have them do unto you" is like squaring a circle—it cannot be done. The Exodus verse conveys one of those unresolved contradictions in the Judeo-Christian myths of Godhead that led Jung to write *Answer to Job*. The theological quandary Jung addressed raises an often ignored but increasingly urgent ethical question: Are we to be judged as members of a

group or as individuals? Western ethical theory, although not the practice, has judged immoral punishment or condemnation of individuals for the "sin" of group membership—at least until Marx and the proto totalitarian opportunists who followed in his wake. Marxism reverts to the primitive ethic of the Pentateuch where individuals can be condemned and scapegoated for the past sins of members of their group. This reversion was far older than Marx, no true innovator. Making group identity paramount has long provided the underlying rationalization for anti-Semitism and various strains of racism, as well as a justification for slavery, segregation, and in our day the bias against LGBTs.

Marx's dubious innovation was to employ the ancient inequity to rationalize violence by the proletariat against the bourgeoisie. Today group sin and group victimhood provide a rationale for the various "affirmative action" and "diversity" schemes increasingly prevalent in the west, though the chief credit for initiating these social engineering machinations belongs to Joseph Stalin. In practice such schemes violate the rights of underprivileged members of groups designated as privileged for the benefit of privileged, politically connected members of the officially designated "victim" groups. Group sin provides a cover for reviving anti-Semitism and for the fashionable scapegoating of white males and Western culture itself for all the sins in world history and all the defects of modern society. Condemning and/or discriminating against individuals to punish groups they belong to violates the Judeo-Christian ethic as surely as slavery or genocide. Yet, as in the days of black slavery, native American genocide, and the rise of Nazism, ethical principles are again routinely discarded in the service of financial gain, political expediency, and social crazes.

6. DECONSTRUCTING OUR HERITAGE. The self-styled intelligentsia of the advanced Western nations, which holds sway over their media, education, and government, seems bent on "deconstructing" and debunking the Western cultural heritage. It seeks to trade the Judeo-Christian ethic for a Marxist pseudo ethic masking ruthless opportunism. Marx was born two hundred years ago, but the fashionable post-soviet Marxism of our media, schools, universities and career bureaucrats and politicians is a relatively new phenomenon. Several factors account for its recent and rapid rise. 1.) The fall of the Soviet Union made Marxism a safer option, since the new Marxists could no longer be accused of aiding foreign enemies. 2.) The civil rights movement and affirmative action gave Marxist free riders opportunities to poach a moral legitimacy they lacked. Charging racism became their ballista for badgering any opponent. White male scapegoats replaced rich bourgeoisie scapegoats for the post-soviet Marxists, many of whom are rich white male leftists. 3) They exploited moral suspicions about churches that were eager to condemn LGBTs and ban contraception along with all abortion, but slow to censure racism, Nazism, and pedophilia in their own clergy. Church morality seemed increasingly spurious which left a moral and ethical vacuum for the post-soviet Marxists to fill with their demagoguery of identity politics. By taking politically correct deconstruction to extremes, they threaten to cut off the West from its taproot myths and its archetypal Godhead whose ethical values have nourished, guided, and sustained the West for nearly two millennia. The rich intellectual, aesthetic, and scientific culture of the

West could then wither, and take with it humanity's most dependable guide.

7. AI IN OUR FUTURE? By 2100 technological advance will have propelled humanity into an entirely new timeframe more different from today than today is from 1900. How many people will be able to compete for productive, let alone creative, roles in our brave new economy without an AI upgrade either through electronics, mechanics, or gene therapy? The personal computer and the smart phone are the beginnings of a series of such upgrades. How many who started as human flesh will need or desire to transform themselves into AI hybrids, or cyborgs, with far greater range of movement and a wider range and depth of talents just to remain useful and productive? The changes will be more dramatic and difficult than just adding a new application to our smart phones. AI poses four critical questions. First, can we make AI an effective tool for facilitating self-realization? Second who will the Transformed be? Third, what happens to those left behind---if any chose or are forced to be left behind? Fourth, will inflexible religious and ideological dogmatism prevent or delay our optimizing the potential of A-1 for human welfare and/or lead into terrifying abuses?

The ways life creates value and meaning across time give us reasons to believe that humans as we know them today, beings with complex feelings and character arising from experience making and living with their decisions, will in crucial respects remain within the transformed AI-human hybrids. The features of humanity that are unlikely to be duplicated fully and effectively in non-organic processes would be in experienced based values and feelings along with intuition, judgment, and character. These features develop from living in an organic body directly subject to the tempo, rhythms, seasons, vagaries, hazards, and occasions of time.

In many respects humans are not very different from the higher apes, we are a big advance mainly in cognitive abilities. Yet the extension of humanity into AI human-electronic beings where electrical or quantum activity replaces biochemical nerve impulses might be the most significant evolutionary advance since plants and animals left the sea to colonize the land. Why so?--because it could make practical the colonization of space. Our bodies, so well adapted to our earthly cocoon, will not get far in space. Interstellar travel cannot be done on a large scale, if it can be done at all. It will be far easier to colonize Antarctica, the Sahara, and our shallow sea beds than any place in our solar system beyond of the Earth. Our prospects will change radically if much of what is essential in the human mind, psyche, and even our personalities could be transferred into AI entities. Such trans-human A1-hybrids might not be a new, alien race any more than Plato, Leonardo, or Shakespeare represent a new alien race compared to savages living before the dawn of civilization or even the groundlings in the Globe Theater. Each evolutionary advance carries over most of its predecessor stage with a few crucial alterations. Evolution moves in a spiraling progression; it never throws away the genetic blueprints or starts over from scratch. Evolution is the great remodeler.

Perhaps mankind can become a race of demigods, like these three supreme geniuses. Our lives before transformation may become analogous to our lives in the womb in preparation for birth. Humanity will survive in AI form because the "feeling-intuition-values-character" part of our animal nature will not become obsolete. It provides the motivation, direction, and integrity essential to judgment and continuity of identity: it provides a reason to be and the motivation to struggle on. Indeed, consciousness itself appears to depend upon and be sustained by our "feeling-intuition-values-character" faculty. Without that faculty there is no reason to be, and, perhaps more important, no reason to change or adapt or even to think, imagine, or dream.

Some individuals may choose to become fully A1 trans-humans who can turn on or off at will, and have no compelling needs to regularly sleep, eat, interact, reproduce, or to die. Each awakening might become a voluntary choice to set forth in a new incarnation, or a re-incarnation if you will. Before a new incarnation we might need and want to live a very long time, experience the entire duration of an ordinary human lifespan, somewhat as George Bernard Shaw envisioned in *Back to Methuselah.*

Time itself makes these dramatic changes possible, and likely inevitable. A fully effective AI-hybrid may need to incorporate the range of human experience, suffering and defeat through achievement and joy, over time. He/she would need ample time to attain stable, enhanced character and ethical motivation in the kind of slow process, like fine wine aging, that life provides. To create value, character, and selfhood, what Jung called individuation, involves the "slow cooking" we see in the evolution of life, of advanced cultures, and of highly individuated people.

Religious and ideological resistance to A-1 and related biological advances poses the danger of going to one extreme and impeding beneficial research and technologies. The controversies over stem cell and embryonic research are harbingers of potentially more serious problems. Medical advances that might save or improve millions of lives could be delayed for years or even decades by the politicization of religious dogmas on the immortality of the soul whose basis is understood by the few yet blindly accepted by the many. Irrational resistance could also blind us to some of the real dangers of genetic and other experimentation, or it could drive such research into the shadows within countries that lack adequate safeguards against abuses by people with copious resources but scant regard for ethics and the welfare of humanity. Unbiased, clear headed thinking is essential for realizing the benefits and avoiding the pitfalls of A-1 and genetic engineering.

8. The subtle savagery of INSTITUTIONAL PROTECTIONISM is the eighth dire challenge and the eighth unrecognized deadly sin. It is our gravest threat to freedom and most serious impediment to accountability, reform, and progress. Its abuse is the royal road to totalitarianism. One of civilized man's most enduring problems, it never seems to grow old but always renews itself as we invent new institutions and expand old ones. Institutional protectionism has afflicted the military as long as there has been a military. It plagued the churches from almost their beginnings and today has brought the Catholic Church shame and disrepute over its systematic cover ups of pedophilia among their clergy. Far from curbing the excesses of institutional protectionism, all religions encourage them whenever their own interests are at stake. One wonders

if Whitehead might have had their ruthless self-protection in mind when he remarked "Religion is the last refuge of human savagery." Whenever institutions are not effectively held accountable to outside standards and oversight, abuse and corruption are free to thrive. More evil is done in the name of protecting institutions from accountability than is committed out of lust, greed, pride or any of the other seven deadly sins.

Institutional protection provides a rationale for suspending ethics, basic decency, and common sense that few dare challenge. 'Do unto others,' unless of course that might impinge upon the interests of the institution that employs you, then it's either forget the golden rule or forget about keeping your position and very likely your livelihood. Forget about honesty and truth as well, because you must never say anything that reflects poorly on your institution no matter how badly it behaves or performs. Memorable examples appear in the recent television miniseries *The Americans* where otherwise ethical and sympathetic people working for the FBI and the KGB routinely commit grossly unethical and even cruel, murderous acts in service of their agencies. To turn from fiction to reality, the current political controversies in the real FBI and DOJ provide fulsome examples of abuses at the top, but zero examples so far of either whistleblowers or anyone resigning to protest the abuses. The corrupting demand for putting institutions above ethics, honesty, and accountability becomes more pervasive with the rising power and influence of all types of bureaucracies. Rampant urbanization and population growth spur the bureaucratization of modern life which results in a dumbing down of personal ethics and abrogation of responsibility. We have passed 1984, yet a more affluent version of George Orwell's nightmare threatens to be building in the future of China, Russia, Europe and even the USA.

If we are to curb institutional protectionism, society must regard serious ethical abuses committed to protect institutions as morally reprehensible, and legally criminal like murder, rape, and genocidal hate crimes. The penalties for it must be swift and sure. The current practice, however, is to punish whistle blowers for exposing institutional abuses and reward those who cover up the abuses—they are deemed "trustworthy." This is par for the course in a totalitarian system, but it corrupts, subverts, and can eventually destroy a free society.

Jesus never let tact temper truth with the Pharisees, and he was wonderfully clear on the bounds of institutional loyalty: Render to Caesar the things that are Caesar's; and to God the things that are God's." Matthew 22:2. There are limits to the things that are Caesar's, and Caesar's things do not trump the things owed to God, chief among them adherence to the Golden Rule. Jesus well understood that those who put loyalty to their institutions above duty to God make the institutions their god. Alas, St. Paul muddied the waters, as he was wont to do: "Let every person be in subjection to the governing authorities. For there is no authority except from God and those which exist are established by God," Romans 13:1. Rooted therein is the pernicious notion that the Church, the state, and other governing authorities are sanctioned by a God who resembles a king, that is an outsized ego. Putatively sanctioned by God, Christian kings and popes acted as if they were deities with authority to curtail or suspend the Golden Rule. Divine right monarchy has seen its day, but the monarch's thrones have been occupied by political opportunists attended by vast bureaucracies ever ready to sacrifice individual rights on the alters of institutional protectionism.

MEDITATION XXIV: LIVING WITH THE GOD IN TIME

Actual God, The God in Time, made, by the irreversibility of time, a universe where suffering, injustice, irreparable loss, tragedy and death are inevitable. Yet the creative nature of time makes order, truth, value, beauty, enjoyment, and meaning not only possible, but evident to anyone. Indeed, the irreversibility of time and the interconnectedness and interdependence of everything in time give Actual God moral qualities and provide a universal ground for ethics. The human condition affords eight ethical/moral guidelines for optimizing the good and overcoming evils by learning to live in harmony with The God in Time.

1.Only adapt. To survive and thrive we must adapt. Adaptation is the first law of life and the prime driver of evolution. Adaptation requires humility, courage, consciousness, honesty, and resourcefulness. We need humility to admit that our beliefs or practices have proved wrong or outdated. We need consciousness and honesty to see things as they are. We need courage and resourcefulness to change and adapt. Arnold Toynbee called it challenge and response, noting that the responses of civilizations to challenges shape their history and often determine their very survival. Civilizations, institutions, and individuals all become enamored of comfortable, flattering, if unrealistic, self-images. The collective ego identifies with these images. Thereby, the voice of the larger self and the capacity to adapt to ever changing, irreversible time can be seriously muted or lost. We are above all else beings who must adapt consciously, who can freely choose what we become from the options that our place in time offers. Our enhanced consciousness makes us very different from creatures with more specific, instinctive adaptations tailored to specific locals and situations.

2.Only connect. Discernment of relationships is the heart of consciousness. Reality, as we know it, is a continuum of events interrelated in complex temporal processes. All is connected, all things are interwoven, all is interdependent. To adapt, survive, and thrive we must connect. We must understand our intrinsic interconnectedness and interdependence with nature, our civilization, society, and personal circumstances. Our most challenging quest for connection is inside ourselves; we must connect or integrate the parts of the psyche to achieve individuation. The scapegoating endemic to anachronistic religion reflects and creates inner division, or alienation from parts of ourselves. Religion is that which binds, and so it should be the grand facilitator of connection. Instead, religious anthropomorphism brings connection's opposite, an alienation from nature that leaves its followers blind to some of society's most urgent challenges, like protecting the biosphere and other species. Equally bad, the alienation blinds us to our interdependence with the self and with the rest of humanity. It impedes self-knowledge which in turn impairs our ability to adapt and connect.

3.Love requires truth and truthfulness—a simple lesson that King Lear and many others had to learn the hard way. We do not lie to, mislead, or deceive those we truly love, and we must listen to them as we must listen to our own deepest self. Successful adaptation requires truthfulness as does genuine connection with others. The first and hardest task is being truthful and faithful to ourselves. Truly loving the self requires the ego to listen to the self which is a rare and difficult art. Truly loving others requires listening to them, which is a multifaceted and almost equally difficult art. The rewards of truth to the self and others are authenticity and integrity, without them love becomes a hollow passion.

4.Correct injustice and falsehood. Life hands us lessons; it also hands us duties and tasks. It is the duty and task of all to strive to correct the injustices we suffer and commit, and to counter falsehood with truth. Injustices and falsehoods are social malignancies, if left alone they metastasize like cancers into the body of society. The victims of injustice are the people best suited to correct injustice because they have the most authentic motivation and understand it best. Remaining silent, they become accomplices to its spread. Those who know the truth are best able to correct and stop lies. Injustice and falsehood corrupt and destroy love, faith, and trust. Love, faith, and trust sustain society. As they decline so does a society or a nation. All great social disasters, like wars and economic collapses, have their roots in injustices and falsehoods that have festered uncorrected.

5.Know that things fall apart, but also that things coalesce. Entropy and coalescence are two of time's elemental principles for life and the universe. Things decline gradually into disorder, people and animals age, plants fade, stars burn out, atoms decay etc. All the while other things coalesce into novel wholes that are more intricately organized than their constituent parts: a chaotic stellar dust cloud becomes a magnificent solar system like our Sun's; heavier elements form in the heart of stars made of hydrogen; these elements can eventually coalesce into complex compounds from which life emerges; magnificent birds evolve from cold blooded creatures; and mankind and tigers evolve from rodents. In human society the greatest force of coalescence is love. Jesus is its prophet and Shakespeare its poet.

While things decay and deteriorate from their past states, in the future other things will coalesce into more advanced orders that are rich in novelty and beauty—such is the vast, spiraling creativity of time. Jung uses the term constellate to refer to a manifesting of events with related meaning that is precipitated by archetypes active in the self's drive toward wholeness. Whitehead speaks of concrescence, Teilhard of complexification, Bergson of creative evolution. These are analogous perspectives on the organizing principle of our interdependent, creative universe which I refer to as coalescence and whose opposite is entropy. Life is about coalescence. Creative evolution, coalescence, concrescence, and complexification drive life forward. Time's basic duality of coalescence and entropy runs through all that moves, grows, and changes. We must open ourselves to the challenges of life and be ready to venture forth when a new world beckons. Every new day presents creative possibilities along with the perils of alienation leading to entropy. The future will always be fraught with novelty in creation and decay.

6.Learn to listen to the singing of the soul and move with the music of time. Time has its own music, its complex structures and patterns, a verity Shakespeare stressed. There are small eddies and crosscurrents as well as mighty tides in the affairs of men, we must listen to the music of time to navigate them. Those who ignore the strains of time, the peculiar rhythms, harmonies and disharmonies of events small and great, cut themselves off from the flow of life, lose its direction and meaning, and miss its beauty, truth, dangers, and joys. The soul sings to us not just of what is past, passing, and to come but of what is real and true, warning against the false and shallow. Integrity of life is the reward of those who listen to and heed the songs of their soul.

7.Welcome adventure, expect the unexpected. We grow, learn and advance through adventure. Every true adventure involves encounter with the unexpected, a truth dramatized repeatedly in the greatest of all adventure stories, Homer's *Odyssey,* as well as by the runner-ups, *Don Quixote, Gulliver's Travels*, and *Moby Dick.* Creativity is a necessary response to the unexpected. It can awaken, enlighten, refresh, and renew individuals who have grown lethargic and societies that are stagnating: without creativity both age and die. Creativity is Time's gift to adventurous spirits, like the bold Ulysses, who have the resourcefulness to adapt and connect and the courage to take a chance.

8.Give and hazard all—. Kierkegaard called for a leap of faith, but that was a retreat into a re-affirmation of past beliefs many had come to doubt. Jung called for *pistis,* a type of faith that contrasts to blind submission to an external authority or willed belief in doctrine. *Pistis* refers to a conscious trust in the creative, sustaining abilities of the self, or, to use Jung's words, "fidelity to the law of one's own being." Shakespeare's phrase from *The Merchant of Venice* enjoins a similar faith, an enterprising and adventurous leap into an unknown future, which requires the courage to give and hazard all to create something novel and valuable. We must dare to create and venture into new paths, not stubbornly or sentimentally cling to past ways that no longer work well.

Shakespeare mirrors the wisdom of Jesus's parable of the stewards, never rest content with just protecting what we are given, we must use it to improve our lives and the lives of others. Avoiding evil was insufficient for Jesus, instead we must adapt to life's challenges and opportunities by risking all we have to create new good. The parable embodies the ethical genius that sets Jesus apart from and above others content to ask for mere conformity to social norms. It is an argument for his divinity, and a reason for the success of Christian civilization. Adaptation itself ultimately requires connecting, opening our eyes, moving with the rhythms of time, and the courage not to believe so much as to take the chance of failing in order to discover, innovate, and create. Whitehead says that God Himself gives up absolute power to make his creatures free, and so He hazards the risk of failure in an act of supreme love and creativity. Nothing new and good is ever created without risk of failure. Hazarding failure to create new good is the most heroic exercise of freedom and shows the highest love.

* * * *

Despite the limitations of ancient times, most of what ordinary men need to know to live a good life can be summed up in three teachings of the founder of Christianity: the lesson of the parable of the stewards, to use whatever you are given to create new good; an ethical rule: "do unto others as you would have them do to you;" and a promise "the truth shall make you free." In short, put your time and gifts to good use, love and respect others, and seek and live by the truth. Add to these the active wisdom of the West and Shakespeare and the contemplative wisdom of the East to get more in touch with the flowing energies of nature, time, the soul, and the self, and you have the essentials human beings need to find value in the world and create meaning in their lives.

What role has man in the universe of The God in Time? If a purpose of our time bound, interwoven, interdependent universe is the creation of value, beauty, truth, and meaning, mankind can and does partake of that. In my garden I have a cactus that flowers for a day once a year. Its flowers are no less glorious and meaningful for their brevity. So it is with individual human lives and human civilizations on the time scale of the universe. Art, Shakespeare tells us, and he and countless other artists demonstrate, can capture the moments of glory, meaning, brilliance, and beauty preserving them for subsequent generations. Is the race of man like a cactus flower? Perhaps so, yet that does not make us meaningless. In this universe value and meaning lie in the creation of multitudinous forms of beauty and the discovery of truth which unite us with all life in all time. For mankind love is the most beautiful and meaningful thing we can create, and truth the most useful and valuable thing we can discover. In loving we unite in a timeless moment with our ultimate creator, The God in Time. Fortunately, the creator has not limited us to romantic love. We can love many things including all those that possess the truth of beauty and the beauty of truth.

JAMES DRISCOLL

A PHILOSOPHICAL AND LITERARY GLOSSARY OF TERMS AND CONCEPTS

Jung created an entire vocabulary of psychological terms and concepts that allows us to more easily say and understand and define things that previously were difficult to articulate. Unlike most glossaries of Jungian terms, which are clinically oriented, this glossary is tailored to literary criticism, religion, and philosophy. Many clinical terms, like neurosis, are omitted while other non-Jungian religious and philosophical terms are included. The focus is on the essential terms and fundamental concepts employed within this work. Readers not well versed in Jung may profit by initially perusing all the entries and referring back from time to time. Readers with broad knowledge of Jung may nonetheless want to review key concepts and especially the entries on criticism and religion. Many of the definitions reflect non-Jungian insights, like those of A.N. Whitehead, Henri Bergson, Heidegger, and feminist scholars. The glossary as a whole sketches the conceptual underpinnings of *The God in Time.*

ACTUAL GOD. Jung uses the term to refer to God as a metaphysical entity. My definition of Actual God: *the dynamism that is and moves through time and by that means creates the universe.* Actual

God entails and sustains the creative interconnectedness and interdependence of everything in time, and thereby integrates the universe and maintains homeostasis. In coalescing the *unus mundus*, and through Time's irreversibility, Actual God gives order, value and meaning to all things within time and the universe; It thus acts as a metaphysical guarantor of the meaning of life and the validity of religious experience. Actual God is inherent in the fundamental character/principle of time: irreversibility and its corollaries order, value, meaning, creativity, adventure, novelty, responsibility, and universal interdependence. The irreversibility of time and the interconnectedness and interdependence of everything in time give Actual God moral qualities and provide a universal ground for ethics and aesthetics.

ADAPTATION. To survive and thrive we must adapt. Adaptation is the first law of life and the prime driver of evolution. Adaptation requires humility, courage, consciousness, honesty, and resourcefulness. We need humility to admit that our beliefs or practices have proved wrong or outdated. We need consciousness and honesty to see things as they are. We are above all else beings who must adapt consciously, who can freely choose what we become from the options that our place in time offers. Adaptation is the great driver of evolution. Jung contrasted adaptation to conformism, and saw failure to adapt as one feature of neurosis or psychosis, which often appears when the psyche faces intolerable suffering or injustice. King Lear's madness when he flees onto the heath in the storm is an example.

ADVENTURE is a key concept for Whitehead. We grow, learn, and advance through adventure. Every true adventure encounters the unexpected, a truth dramatized repeatedly in the foremost adventure story, Homer's *Odyssey*. Adaptation through creativity is an essential response to the unexpected. Creativity is Time's gift to adventurous spirits. Creativity sustains the self, but without adventure creativity atrophies along with the self. Progress and innovation begin with and require adventure.

ALCHEMY. Jung believed that alchemists projected psychic contents into matter and thereby employed alchemy to work indirectly on the psyche which made alchemy a precursor of modern psychology. Hence, alchemy forms a repository of evidence about and insight into psychic processes, particularly individuation. Like art and analytic psychology, alchemy tends to be compensate for the deficiencies of the cultural canon. The alchemists arranged the elements used in their experiments in pairs of opposites, their objective being to make from the conjunction of opposites new elements. Just as alchemy sought to differentiate the elements through work with opposites, so Jung's psychotherapy seeks to differentiate or individuate psyche through work with opposites. The goal of alchemy, the philosopher's stone, symbolizes the goal of therapy, wholeness. Art, especially painting, has an alchemy all its own, and great art seeks the alchemists goal of wholeness.

ALIENATION. While currently associated more with Marxian than Jungian thought, this fundamental concept goes back to Augustine and Greek tragedy and is critical in Shakespeare and Milton. Alienation, dependence, and interdependence define the spectrum of possible relationships of individual to God and of ego to self. In traditional views that render God wholly other, sin is a metaphor for the alienation from God that exists in the cosmos. In Marxian ideologies alienation from historical necessity and the collective induces betrayal or bad faith, the Marxian equivalent of sin. For Jung initial alienation of ego from unconscious oneness is necessary to individuation, but permanent alienation of ego from self and *unus mundus* generates a sickness of soul whose symptoms are despair and meaninglessness. Alienation and inflation, as the internal and external facets of uroboric regression, are symbiotic evils. Alienation feeds inflation, especially in the forms of pride, scorn and envy, and inflation always alienates others: the presence of alienation and inflation ever blocks or stifles love. The remedy to alienation and inflation is love or interdependence with others, self, and *unus mundus.*

AMPLIFICATION is a method of interpreting symbols, dreams, and myths that clarifies their meaning makes it more comprehensive by drawing upon their cultural and mythological context. Consider how Greek culture and mythology can amplify the Oedipus myth. The gods, the Greeks believed, are the driving forces of myths; Jung would tell us that gods represent archetypes and the gods in each myth tell us which archetypes are active. The active god-archetype driving the Oedipus myth is Poseidon. Poseidon carries archetypal feeling, as his brother Zeus carries archetypal thinking. The Oedipus myth began not with Oedipus's crime but earlier with Poseidon as erastes to mortal Pelops' eremenos. Afterwards, when the mature Pelops needed help securing a bride,

Poseidon, remembering his joy in Pelops, came to the man's aid: thus Poseidon and Pelops established an ideal erastes-eremenos relationship. The offspring of Pelops's marriage was Chrysipus. Laius, visiting Pelops, lusted for Chrysipus, stole the boy by force, and violated him, causing his death. By disregard of feeling, Laius became a destructive erastes contrasting with Poseidon's beneficial erastes. The disasters that befell the house of Laius through his offspring Oedipus, who symbolizes thought divorced from feeling, were retribution of Poseidon (feeling) against Laius's unfeeling violence. Thus, Jungian amplification yields meanings in the Oedipus story beyond what Freud found: the rivalry of father and son for the mother is a destructive effect of the damage to the feeling function that results when erastes-eremenos male bonding is violated. Such amplification can be applied to other important literary works, among them John Milton's major poems, and many of Shakespeare's plays. In *King Lear* amplification of the stages Lear passes through and of the figure of Cordelia reveal the archetypes of the true Quaternal Christian Godhead. Although *King Lear* deals fundamentally with an archetypal situation critics tend to look for psychological realism. Amplification avoids loggerheads that result when we seek a realistic explanation of Lear's initial unconscious, irrational and tyrannical behavior with his daughters by showing that the early Lear is driven by the Yahweh archetype and is at odds with the *anima.* Betrayed by the daughters he trusted and losing all, Lear on the heath in the storm embodies Jung's Job archetype and by extension the crucified Christ. His final defiance manifests the Promethean archetype. While Cordelia, embodying truth and love, manifests the Paraclete or wisewoman archetype. Amplifying the archetypes shows Lear's story to be a dramatization of the dynamics of the Judeo-Christian Godhead. Just as the meanings of Oedipus story are enriched and deepened by amplifying archetypes, amplification expands the richness and depth of meaning in *King Lear*.

ANIMA. In Jungian psychology the term has two distinct meanings. A*nima* refers to the contrasexual element in the male psyche that, appearing as a single figure, plays mediatrix to the collective unconscious. However, *anima* also means soul, and as soul it carries the eros, or love function, that involves us with others, with the inner, imaginal life of the psyche, and with nature. Carrying the eros function, *anima* appears in tandem or coupled with *animus* to form the androgynous syzygy or the divine androgyne a personification of male and female held in balance. In Shakespeare Prospero and Cordelia embody this sexual balance, and to a less developed extent Viola in *Twelfth Night* and Portia in *The Merchant of Venice*. Jung believed that the historical figure of Jesus achieved androgynous psychic wholeness. Some leading *anima* figures are Aphrodite, Helen of Troy, Mary, Dante's Beatrice, Shakespeare's Juliet, Rosalind and Cleopatra, Milton's Eve, and Joyce's Molly Bloom.

However, imbued with patriarchal values, Jung at times blamed the contrasexual "inner woman" or *anima* for men's weaknesses. Such scapegoating implicitly bolstered the myth of male superiority and inhibited recognition that both men and women have *anima* in the sense of soul just as both have egos. In either sex ego tends to behave in an active, assertive manner often conventionally identified as masculine. While the protean *anima* is technically androgynous, in Western culture its subtle, "feminine," nurturing qualities tend to predominate. The differences between the two sexes on *anima* and *animus* are differences of identification: the male ego usually identifies with *animus* or spirit and feels separate from anima or soul, while the female ego often associates with *anima* and dissociates from *animus*. Possession by either *anima* or *animus*, according to Jung, causes over expression of traits characteristic of the opposite sex resulting in effeminate, sentimental men and bossy, opinionated women.

Taken together *anima* and *animus* form the yin and the yang, the eros and logos, of androgynous wholeness. *Anima* gives soul, *animus* gives spirit, and shadow gives substance in both sexes and all sexual orientations. In either sex *anima* supplies the guiding, healing, nourishing imagination that puts us in touch with the inner world of myth, dream, and fantasy, and with nature. When *anima* is repressed, intellect, deprived of imaginal nourishment, becomes desiccated and shallow. The muse, that inspired guide and voice of imagination, is *anima* herself. Feeling and especially intuition tend to fall within the domain of *anima,* while thinking and sensation fall more in the realm of *animus*.

ANIMUS. As *anima* embodies soul and *eros*, so *animus* embodies spirit and *logos*. *Animus* in its general sense, refers to the energetic, independent spirit present in either sex, most commonly as assertiveness, initiative, and adventurousness and curiosity or the exploratory drive. Being independent, *animus* functions differently in every individual. Jung, however, made *animus* the contrasexual part of the female psyche and chauvinistically associated it with a stereotyped willful opinionatedness. Jung's view of *animus* fails to recognize that patriarchal repression of logos or thinking in women drives *animus* into the shadow giving it the negative qualities he mistakenly attributes to *animus* itself. If women freely live out logos-*animus*, it will not become an exaggerated parody of a man, it will be integrated as a constructive part of the feminine ego. Possession by an *animus* or *anima* that has been repressed into shadow, however, gives a person the negative qualities associated with the opposite sex. Such is the case with Milton's Eve when, deceived by the serpent personifying her repressed *animus* grown insubordinate, she strives for mastery over Adam. Lady Macbeth is Shakespeare's prime instance of malefic *animus* possession. Other examples are Lear's evil daughters Goneril, and Regan, and the witch Sycorax in the *Tempes*t.

In contrast to the contrasexual *anima* of man, with which it stands paired in the syzygy, the woman's *animus* appears not as a single figure but as a multivoiced group. Sexual orientation, no less than sex, influences the ways *anim*a and *animus* are manifest in individuals. As heterosexual men tend to project *anima* on women, so homosexual men tend to project *animus*, in the form of energy, talent, and initiative, on men they admire. While heterosexual women usually project *animus* on admired men, lesbians often project both *animus* and *anima* on admired women. *Animus* and *anima*, or spirit and soul, can be viewed as complementary aspects of a single archetype of androgyny; accordingly, in the original Latin the two terms were used interchangeably. Both aspects are present in each sex and sexual orientation, but the degree of conscious identification or projection varies with individuals.

ARCHETYPES are inherent possibilities of psychic functioning and structuring derived from inherited neuropsychic structures and based on situations common to humanity. Although they are not drives or instincts, they predispose the mind to formulate certain types of ideas and images in common situations, even as instincts unconsciously predispose animals to certain types of behavior. They come in two principal forms: (1) personified archetypes such as *anima*, shadow, the mother, and the wiseman; (2) archetypes of transformation, which involve a basic situation or pattern like birth, *conjunctio,* initiation, marriage, quest, and death. Though archetypes appear embedded in images, they themselves are not images but psychic predispositions to form and organize images. Archetypes differ from animal instincts in that men, having the gift of reflection, can consciously oppose, tame, direct, and transform archetypal energies. Furthermore, archetypes combine and interface so that manifestations of particular archetypes are always modified and shaped by the other archetypes constellated in the individual's psyche and in his social situation or cultural milieu. For example, the shadow, as other archetypes influence it, will become Satanic in one manifestation, Promethean in a second, Mercurial in a third, and merely clownish in a fourth. The archetypes broad range of reference makes uniform formulation impracticable.

Only a limited number of archetypes are constellated in a person at any one time; the rest remain potentials of the collective unconscious. The psyche tends to constellate those archetypes that compensate for conscious imbalances or deficiencies. Its purpose in constellating an archetype is guidance and transformation. When archetypes are constellated, in a person, a society or a civilization, psychic energy flows to them making them powerful ordering forces like magnetic fields whose effects can be transformative. At times archetypes can become so potent they gain a will of their own, which can take possession of ego itself or an society. Balance returns as we become conscious of the archetype's power.

Specific manifestations of archetypes, whether historical such as Jesus and Buddha, mythic like Lucifer and Dionysus, or conceptual like the One of Plotinus and the Logos of orthodoxy, never exhaust the possibilities of the archetypes because archetypes always transcend their manifestations. Finally, archetypes are more like Kantian categories than like Platonic forms: being numinous and vital psychic forces, they transcend and generate our conscious ideas.

ART is a means the self employs to send forth imagistic messages through *anima* or sometimes *animus*. Anima's role as channel for images explains why the poet's muse is feminine. The types of written art that most readily serve the self's purposes are primitive myths, fairy tales, and visionary literature, the art that resembles dreams. Here it is often more rewarding to approach characters as archetypes than as real people. Wherever the archetypal element is pronounced, even in exceptionally complex works like *Paradise Lost* and *King Lear,* this observation holds true. Archetypes in art always compensate one-sidedness in the conscious attitude or cultural canon in order to advance wholeness. Art is, as Jung and Whitehead understood, one of the self's great teaching and healing tools. Whitehead saw God as the poet or artist of the world creating truth, beauty, and goodness in time.

COALESCENCE. Just as things fall apart because of entropy, some things, such as organic entities, have a tendency to coalesce into systems of increased order and stability. Whitehead said, "It lies in the nature of things that the many enter into complex unity." Concrescence or coalescence or complexification would be, then, the universal engine of creativity that continuously brings novelty to birth to balance destruction. Things coalesce into novel wholes that are more complex and intricately organized than their constituent parts: a chaotic stellar dust cloud becomes a magnificent solar system like our Sun's; heavier elements form in the heart of stars made of hydrogen; these elements can eventually coalesce into complex compounds from which life can emerge; primitive birds evolve from cold blooded reptiles; and mankind and tigers evolve from rodents. Similarly, while they age, that is fall subject to physical entropy, humans through individuation have opportunities to expand consciousness or coalesce toward a greater psychic order that creates novel wholeness and builds soul—e.g. Prospero and the wisdom of age. Coalescence might eventually avert entropy's heat death of the universe. It has billions of years to find ways to do so. Coalescence is essential to the formation of godheads, as the dominant ideas, values, and archetypes of a culture or civilization tend to coalesce into a Godhead. In human society the greatest force of coalescence is love. Jesus is love's foremost prophet and Shakespeare its supreme poet.

COINCIDENTIA OPPOSITORUM (sometimes *complexio oppositorum). *The principle that opposites or polarities contain, generate, and eventually join each other, is fundamental to Jung's thought from his *Red Book* forward. Recognizing *coincidentia oppositorum* or interdependence of opposites, Jung theorized, enables us to cope with duality, resolve divisive conflicts, and resist one-sided ego commitments. *Coincidentia oppositorum* entails that attributing a quality to God (or any other entity) necessarily posits its opposite: i.e., you cannot make God light without giving him his dark side as well. *Coincidentia oppositorum* is closely related to *unio oppositorum,* both terms supplying alternatives to the *summum bonum* doctrine of deity. *Coincidentia oppositorum* and *unio oppositorum,* Jung held, explain the dynamic nature of the Deity manifest in its power to transform conflicting opposites into a synthesis or homeostasis that preserves their essential tension and energy yet generates new meaning. Unlike the *summum bonum, coincidentia oppositorum* can account for evil without the pitfalls of dualism, of blaming man for all evil, or of denying evil's reality. Jung held, moreover, that understanding and accepting the principle of *coincidentia oppositorum* could encourage integration of shadow and discourage the hypocrisy, scapegoating, and alienation that occur when we repress and our shadow and then project it on others.

Coincidentia oppositorum represents Jung's psychological and ethical ideals of (l) a balanced wholeness that incorporates opposites while maintaining their vital tensions and (2) an interdependence that ends alienation while upholding respect for the autonomy of others. Jung's ideals contrast to the orthodox monotheistic religions' ideals of a one-sided, willed goodness or purity achieved by splitting opposites apart, embracing one and repressing the other. Unlike willed purity, wholeness and interdependence recognize that energies contained in opposites generate the psyche and are necessary to its life, just as the energies in matter generate the physical world. The eros function of the syzygy, or androgynous pairing of *anima* and *animus*, manifests and realizes *coincidentia oppositorum* among the archetypes of the psyche. *Coincidentia oppositorum* means that pairs of archetypal opposites, when truly understood, have a Janus quality in that they become complementary faces of dual archetypes. The reconciliation of opposites is realized through symbols. For example, the reconciliation of the Job and Satan-Prometheus archetypes in the Quaternity is realized symbolically in the Paraclete. This illustrates the power of the psyche, when confronted with seemingly irreconcilable opposites, to create a symbol expressing a third possibility that can show the way out of an apparent deadlock.

Jung cited the first law of thermodynamics which states that energy requires opposing forces. Opposites are essential to the dynamism and dialectic of reality and of the psyche. Shakespeare's corollary to Jung's *coincidentia oppositorum* is the idea of reconciliation of opposites in a metastance. In Shakespeare you can always hear the resonance of opposites. The Shakespearean resonance of opposites shows that the bard, like Jung, was ever aware to the dualities that mark the existential challenges of time. Like Jung also, Shakespeare realized that lasting resolution of the conflict of opposites was an impossibility.

COLLECTIVE UNCONSCIOUS. As distinguished from the personal unconscious, a repository of inherited archetypes common to all humanity. While it comprises much of the total psyche, it is not the whole and should not be confused with the self or that image of self we call God. It contrasts to the collective consciousness or cultural canon which act as the ego and superego of society. Being the seat of the archetypes, the collective unconscious is a vast reservoir of psychic possibilities and divergent energies. It is also the source and rough equivalent of the polytheistic pantheon.

COLLECTIVITY and **COLLECTIVE THINKING** project God upon the collective, which is usually imaged after the father archetype. These modes are evident in the herd behavior and totalitarian mentality common among twentieth-century men (Jung called them "provisional men") and in mass movements, especially religious ones, during previous ages. The protagonist in Albert Camus' *The Stranger* is a literary portrait of this type of person. Such mass behavior and thinking, Jung believed, proves inimical to the individuation process, which is unavoidably an individual struggle. Mitigating against the authentic, individual decisions that form the basis of ethical conduct, collectivity and collective thinking bear a heavy responsibility for the moral and political crisis of modern man and for his spiritual drift and despair. They lie behind psychic epidemics such as wars, persecution, and mass hysteria. The remedies are, Jung insisted, is: 1) to seek God within the self rather than following the collective and projecting God onto church, party, or state; 2) pursue individuation and avoid the pitfalls of conformity. The more individuated a person becomes the more he will diverge from the collective.

COMPENSATION activates repressed or neglected opposites to achieve *coincidentia oppositorum,* establish balance, and enact psychic transformation. Compensation balances, fine tunes, augments and regulates to achieve the homeostasis essential to psychic health. Compensation disrupts the conscious perspective. Whenever a conscious tendency becomes inflated or deflated the psyche surprises the ego by activating compensatory opposites. An archetypal perspective sees events in pairings of compensatory opposites. The tendency of unconscious emergents, in dreams, symbols, visions, and of visionary art, to compensate for one-sided imbalances in conscious attitudes or the cultural canon provides a fundamental guideline for Jungian literary and aesthetic criticism. Similarly, Jung believed that alchemy was compensatory to the deficiencies of established religion and theology.

COMPLEX constitutes a feeling toned group of psychic images usually centered around a core of one or more dominant archetype with a specific emotional tone or affect. Complexes are lodged in the personal unconscious and often arise from the shadow or the *anima-animus*. When complexes split off from the ego and become autonomous, like alternate psyches, they threaten the integrity of the entire personality. A complex may inundate the ego, as in inflation, alienation, deep depression, or psychosis or possession, as with Shakespeare's Iago whose ego is possessed by shadow. Complexes are a refutation to monolithic egocentric theories of human personality. Jung saw four chief complexes operating within the Godhead, the parents (father-mother) the two brothers Christ and Lucifer, and the wise person.

CONCRESCENCE is Whitehead's term for a merging or symbiosis of many discrete elements or entities into a unitary whole. It refers to a process in which many things are subordinated to constitute or create a novel entity. It parallels coalescence and Teilhard de Chardin's complexification, and is analogous to Jung's individuation, transformation, and *conjunctio*.

CONJUNCTIO is an alchemical term for a metaphoric sexual union involving transformation of paired opposites into a new element manifesting a new meaning. Jung saw the conjunctio as the central idea of the alchemical process, and his identified it with the archetype of transformation central to his theory of individuation. The opposites are often *anima* and *animus* and the new element the androgyne. *Conjunctio* symbolizes the process of raising opposites from the unconscious to unite them in consciousness, whereby the psyche advances individuation toward achieves wholeness within itself. Being a transpsychic process, *conjunctio* is not in principle readily amenable to scientific explanation.

CONSCIOUSNESS is that part of the psychic realm that ego perceives and within which it acts. It stands opposed to the unconscious, which remains hidden to ego, except for when it communicates through dreams, synchronicities, and numinous events. Consciousness mediates between the outer and the inner world through the four functions of thinking, feeling, intuition, and sensation. By differentiating opposites, it individuates the psyche. The objective of Jungian analysis is to develop consciousness and bring its unifying center, the ego, into harmony with the greater self. While consciousness is usually associated with ego, the *anima*, the *animus*, the shadow, and other archetypes can acquire their own peculiar consciousness.

Consciousness subsists within time. Time is a necessary condition for conscious thought since without time nothing would be differentiated from anything else and consciousness depends on perceiving relationships and making distinctions between separate things, feelings, ideas, and events. Unconscious contents are undifferentiated as to what belongs to or does not belong to oneself, being undifferentiated they often appear to be outside of time. The light of consciousness illumines time. There is more to consciousness than just its own light, consciousness partakes of all the senses, sight, sound, smell, feeling, and the intangible senses sometimes called intuition or insight. The light of consciousness in other persons, or intelligent beings, also illumines time for each of us; however, it is more like a reflected light than a direct light. Mystic vision, some believe, is the supreme manifestation of the light of consciousness; others might opt for a peak artistic or scientific insights.

Consciousness arises with the discernment of relationships of events within time. All that consciousness can be aware of exists within the fields of time. Consciousness can be defined as a unifying and coordinating force behind the entire interdependent and interrelating processes that form and maintain across time the sense of identity of individual beings. Consciousness both depends on perceiving relationships and makes possible the discernment of relationships, including the intuition of temporal duration as distinct from the mere perception of change. Consciousness reveals how time and its ever altering relationships give events and everything about them value and meaning. Thus, ideas about time and God develop interdependently with consciousness itself.

CONSTELLATE, literally, to shine forth as a star in a constellation shines forth against the dark background of the sky. It is Jung's figurative term for the activation of an archetype against its unconscious background.

COPERNICAN REVOLUTION IN THEOLOGY (JUNG'S). The universe and/or its God(s) operates not like a decision making ego, but like a homeostatic self, Jung surmised. The paradigm shift from ego to self is Jung's Copernican revolution in psychology and theology. These disciplines, along with cosmology, have yet to incorporate the paradigm shift. Western culture is only beginning to grasp its wide net of implications, and not all were fully grasped by Jung. Momentous among these is the implication that Actual God, the homeostatic self of the cosmos manifest in time, God the creator and guarantor or value and meaning, is essentially different from the monotheistic and polytheistic gods nearly all mankind worship which are modeled on the ego and/or related subordinate psychic functions such as the anima, shadow and superego. Moreover, the god image in the self reflects not these subordinate functions or archetypes, but Actual God, the ordering force of the cosmos.

CREATIVITY. Along with the many and the one, creativity is among three fundamental notions in Whitehead's philosophy of organism. Creativity is crucial to process in that it sustains the "ongoingness" of the universe by moving the many to enter a complex and novel unity. For Jung and Bergson creativity is also a key general concept in that it furthers individuation for Jung and drives evolution for Bergson. For Heidegger and many existentialists creativity is an expression of freedom that is essential to authenticity.

CREATIVE CONTINUA is Whitehead's phrase for the temporal process of the universe. The life processes and consciousness itself are key aspects and manifestation of the creative continua. God who unfolds in the creative continua is its supreme manifestation. Its medium is time.

CULTURAL CANON, the sum of consciously inculcated beliefs, dogmas, and authoritative traditions. While maintaining stable cultural identity by resisting change, the canon tends to inhibit adventure and creativity and thereby make a culture ill-adapted and one-sided. For that reason, in a healthy culture, as distinct from a stagnant culture, the cultural canon is challenged by compensatory unconscious emergents, especially through art. In the West, Christian orthodoxy has long been the mainstay of the cultural canon, but challenges from art, philosophy, science and the occult have continuously enriched, deepened, and enlivened Western culture.

DECISIVE IDENTITY, more existential then a Jungian concept, refers to that truth to self, authenticity or integrity, wherein decisions at once flow from and further shape identity so that one consciously decides or chooses, and thus creates, what one becomes. This requires mature character and discipline, unflinching awareness of alternatives and possible consequences, and the courage to conceive of life as a tragedy. Decisive identity makes us accountable for what we become and allows us to realize the mature freedom needed to consciously guide the individuation process. Cordelia, Prospero, the Duchess of Malfi, and Messiah in *Paradise Regained* are paramount literary examples of decisive identity.

DIFFERENTIATION is the means of creation. The God in Time is the great differentiator. Without time there can be no differentiation. Everything and every moment would melt into the great undifferentiated one, a triumph of monism leaving no room for thought, decision, or consciousness. Differentiation is the defining activity of consciousness. Jung uses the term to denote distinguishing parts from whole. Individuation requires differentiation as does authenticity.

DREAMS. Jung saw dreams as compensatory to consciousness, as is art. Shakespeare stressed the importance and meaningfulness of dreams more than any other major writer. "We are such stuff as dreams are made of/ And all our little lives are rounded in a sleep," Prospero declares in *The Tempest. The Midsummer Night's Dream* celebrates dreaming as a metaphor for transformation. Dreams are frequently premonitory as in *Julius Caesar*. For most people, the dream is their main gateway to the unconscious and the realm of the numinous.

DUALISM, resists interdependence of opposites and *coincidentia oppositorom* by making opposites, especially the opposites good and evil, into absolute, irreducible dualities. Orthodox Christianity rejected the explicit dualism of its rival Manicheanism by declaring evil a mere *privatio boni.* Implicit dualism crept back into its theology with the notion of free will, which makes free agents first causes along with God. Ironically, free will was deemed essential to protect God from the accusation of creating evil with its implication that He is not perfectly good. Jung's dismissal of radical free will as a conceit of ego reflects the incompatibility of his thought with dualism. He sought a balance between freedom and determinism, and stressed the limits the psyche places on the freedom of ego. For Jung, opposites touch, conjoin, and even become each other; they are never irreducible or irreconcilable. The function of *anima-animus* is to conjoin and transform opposites and so manifest the self in individuated wholeness. Dualism reflects a psychology that, locked in the ego-shadow stage of development, depends on repression and cannot accept either mediation from *anima-animus* or the individuated wholeness of the self. While Jung borrowed much from Gnosticism, he never embraced its dualistic tendencies.

DURATION. We understand duration intuitively, not intellectually, Bergson noted. Duration is not something we can perceive by the senses, since we directly perceive only the moment which does not contain duration. The sense of duration comes with reflection and the intuitive realization of the continuum or process of passing of time. The intuition of duration forms the basis for memory and our sense of time. As consciousness expands so do our intuitions about duration and the processes that endure, wax, or wane as time goes by. The more individuated our consciousness, the more we understand what endures for us, how, and why.

EGO, for Jung, is the center of consciousness charged with maintaining identity continuity and mediating between conscious and unconscious. Ego is less than and subordinate to the self, the regulator of the entire person. Jung proposed a self-based psychology in contradistinction to the ego-based psychology Freud propounds and Christian orthodoxy, with its hypertrophied ego-God, assumes. The ego, Jung maintained, is only the subject of one's consciousness, whereas the self is the subject of the entire psyche, which includes the unconscious. Ego identifies opposites and withstands their tensions to mediate them through the work of the *transcendent function.*

The ego, the shadow, the *anima, animus*, *persona* and other archetypal functions are a small archipelago of consciousness in the vast ocean of the unconscious. Ego, as the active center of will and identity, constructs the *persona*, stands paired with *anima-animus*, possesses a shadow, and must cope with the social world, the cultural canon, and collective morality. Standing between the self and its collective shadow, the uroboros, ego must learn to follow self toward individuated wholeness and resist uroboros and its temptation to withdraw into inflated alienation. Among women in Western culture ego tends to be more diffuse and identified with eros; whereas among men it is often more focused and identified with logos, *animus,* or spirit. Even as *anima* often acts as a feminine presence in each sex, so ego often acts as a masculine presence in men and sometimes in women. In women, however, ego attaches to one of the principal feminine archetypes: i.e., mother, daughter-virgin, mistress, wisewoman. Being the seat of the will and responsible for maintaining identity in both sexes, ego seeks strength, security, certainty, and a sense of freedom. These goals are ever sacrificed and the ego ever crucified by the self in its quest for individuated wholeness. Though ego strives for security, because of the self it achieves instead individuation through suffering. Hence, ego, desiring to play the hero in a comedy, is fated to a tragic role.

ENANTIODROMIO refers to a tendency for things to change over time into their opposites, an example being the spiritual Christian culture of the Middle Ages becoming the extremely materialistic Western culture of today. Jung derived the term *enantiodromio* from the philosophy of Heraclitus and made it the basis of his principle of compensation and of the *coincidentia oppositorum.*

ENJOYMENT is a principal term and concept for Whitehead. Life, said Whitehead, is enjoyment derived from the past and aimed at the future. The emotion of enjoyment always comes from somewhere else and is going somewhere else. The upward trend in evolution (marked by improved control of a creature's environment) is driven by increased capacity to enjoy intensities of value. God, Whitehead believed, seeks to facilitate his creature's enjoyment through freedom, creativity, and adventure. Enjoyment is a lodestone and a touchstone value for Whitehead, as consciousness and individuation are for Jung.

EPIPHANY, the term originally referred to a manifestation of deity or a revelation from God. Its modern meaning owes much to James Joyce. For Joyce an epiphany is a sudden manifestation of a thing's essential nature or, in Jungian nomenclature, its underlying archetype. Epiphanies, thus, can be regarded as glimpses of archetypes (for the ancients they were encounters with gods). Hence, we can regard as epiphanies Shakespeare the artist's brief glimpses of the true archetypes of Godhead in *King Lear.* These glimpses, taken together, form, like Job's full theophany, signal advances in consciousness of Godhead.

EROS is the principle of relatedness identified with the feeling function and *anima* and contrasted with **LOGOS** and *animus.* Eros is ambiguous and indirect while logos is definite and direct. Both have positive and negative aspects and operate within both sexes. Eros is identified with the maternal and logos with the paternal archetype. Eros and logos are powerful principles within the Quaternity. The son is traditionally identified with logos, while the Paraclete mostly carries the function of eros. Since the Holy Spirit is neuter in orthodoxy, the feminine or eros function is projected outside the Godhead onto the Virgin Mary. Jung identified five forms or stages of eros expression: the biological, sexual, aesthetic, spiritual and *sapientia,* or wisdom, the last being manifest in the Paraclete and the wiseperson archetype.

ETERNITY is not separate from Time; eternity is the spiraling entirety of time. Paradoxically, eternity is the entirety of time and yet it also lies in the moment. As such it cannot be measured or directly experienced by consciousness within time. There are no meanings, thoughts, feelings, values or consciousness in an eternity beyond time, Plato and his Neo-Platonic followers notwithstanding. Such an eternity is a conjecture defined by/against time, not a verifiable experience we can sense or measure, in short, it does not "exist." It functions like a zero in the number system where the numbers are time itself.

ETHIC OF REBELLION, a key concept of Erich Neumann, arises in reaction to the excesses and defects of the old ethic. The egos of rebels, rejecting traditional values and authority, embrace feelings that have been relegated to the shadow. The rebels, however, suffer from the same inflated perfectionism afflicting those committed to the old ethic: for example, while the orthodox insist, in defiance of human experience, that God is perfectly good, the rebels reject him for falling short of perfection. The clash of the old ethic and the ethic of rebellion characterizes the hostile brothers stage of individuation with its dualist archetype of the Son and Satan and its conflicts of ego and shadow, thought and feeling, stasis and flux. These conflicts are transcended through anima-mediated inward vision.

EVIL, the term is ever surrounded with confusion. Confusion notwithstanding, in post-medieval Western culture there are but three principal visions of evil: evil as impurity, the most primitive of the visions; *privatio boni,* or evil as passive deprivation of good; and uroboric evil, or evil as inflated alienation that actively opposes good. These visions support three basic stances toward evil: (1) One can repress it in oneself and project it onto scapegoats-the position of those Christians who teach will to purity and emphasize the eternal damnation of Satan and his incorrigible followers. This stance is compatible with flux philosophy and leads ultimately to dualism. (2) One can, in effect, deny evil reality, the stance of Plotinus, stasis, and the doctrine of *privatio boni.* (3) One can admit that evil is real in the world, oneself, and God, and set to work transforming it into new good infused with consciousness, the stance of process philosophy and of Jung in his doctrine of *coincidentia oppositorom.*

For the third stance evil is not a reason to question God's love but evidence for it: God allows evil so that through love we might co-create with him a new and more conscious good and thereby know the supreme joy of creating consciousness. Evils that we can transform into new good, like the contents of the shadow, are operative evils. Evils that we fail to transform because of uroboric regression, love's opposite, become moral evils. The only absolute evil is the uroboric counterforce to the self's (or

God's) individuation drive to transform operative evil, usually in the forms of shadow and suffering, into a new and more conscious good. While the uroboric counterforce to self, being an absolute, can never achieve a *conjunctio oppositorom* with self, it is not a dualistic evil to be overcome at the end of time. The continuing struggle against it is necessary to realization of the absolute good of individuated consciousness in self and God. The uroboric counterforce is analogous to gravity in significant respects: (1) we are fated to struggle against it; (2) if it triumphs totally, as in a black hole, it becomes anihilatory; (3) without struggling against it we cannot develop strength and substance.

EXISTENCE OF GOD. Does God exist is the second most basic philosophical question; the most basic question is: do I exist? Descartes answered it: I think therefore I am. A corollary is: I feel, I sense meaning and discern value, therefore God, the ground and creator of meaning and value, exists. Problems arise when people want God to be something more than the possibility of value and meaning in their world or of belief in the significance of human life and our power through free will to shape our lives and make a difference in the world. They want to give God an ego like their own, only more powerful. They want to see God as a big brother, or a parent, a mentor or pal who has a special favoring relationship with themselves. For Jung and this work, Actual God is the self of the universe who interfaces with man through each person's self.

Once we acknowledge that we exist, the existence of God becomes a corollary. The real controversy centers on the nature of God, not the existence. A pantheistic God exists because Time exists. Time creates value and meaning through temporal relationships everywhere all the time. That is what time does, and creation of value and meaning is what God does. So the proof of the nature and existence of a pantheistic God of the universe lies in the nature and function of time. However, the pantheistic God is the context for more limited and personal visions of deity prevalent in human cultures, not an alternative to them. The degree of validity of particular cultural visions of deity, such as the Christian, Jewish, or Islamic, remains an open question.

FATHER, YAHWEH is the male aspect of the primal parent, the most rudimentary archetype of the Jungian Quaternity (the Great Mother, the Paraclete, and Mary represent the female aspect). However, it is the Father who determines the basic archetypal patterns in monotheistic cultures, by repressing the mother and daughter, dominating the sons to create oedipal conflicts, and turning the sons against each other (the hostile brothers) through favoritism. Generally speaking, the Father manifests herd thinking, collective values, and animus or spirit. Its Yahweh (or Allah) forms resemble a patriarchal tyrant, or even a demiurge, in whom conscious ego is immature, reflection is lacking, *anima* is rudimentary, opposites have not been fully constellated, and only the sensation function has been differentiated. The archetypes of the Quaternal Godhead unfold across history to compensate for Yahweh's deficient awareness. The early King Lear who disowns Cordelia for failing to flatter him manifests this archetype.

FEELING, thinking, intuition, and sensation comprise the four basic modes or functions of conscious activity. Feeling, a judging activity concerned with values and timing, stands opposite thinking, which judges logically: feeling is associated with *anima* and eros, and thinking with *animus* and logos. Feeling gives substance to value and moral judgments just as sensation gives substance to physical judgments. Suffering and evil develop feeling, which finds its supreme expression in tragedy. To the thinking function life is a comedy; to the feeling function life is a tragedy. In Christian theology, where the thinking function as logos has ruled and feeling has been repressed or sentimentalized, evil and tragedy are repudiated in a final comedic affirmation. But feeling or pathos is personified in Prometheus-Lucifer- Satan and also in the Son as the suffering Christ. In Greek mythology feeling's premier embodiment is feminine: the goddess Aphrodite, her male analogs are Ares, Poseidon, and Dionysius.

FLUX, along with process and stasis, form three basic philosophical positions on time. A flux universe is pluralistic, temporal, irrational, and imperfect. Flux bolsters the Satan-Lucifer-Prometheus and shadow archetypes, the feeling and sensation functions, reductive thought, relativism, alienation, the ethic of rebellion, an ironic outlook, and the dualistic attitude that makes evil an impurity localized in scapegoats. Flux leads to polytheism even as stasis leads to monotheism.

FREEDOM is defined by three concepts: (1) free will, the ego's power to act, or to choose, free of external interference, which is a necessary though not a sufficient condition for (2) responsibility for the consequences of one's acts and 3) mature freedom. Mature freedom or individuated wholeness is a state in which the ego and the self work in harmony, no single archetype dominates, and decisions flow from the whole self. The androgynous wiseperson archetype is constellated in those who achieve mature freedom. Free will alone cannot make one responsible or give decisive identity. These entail at the very least the character maturity to exercise self-discipline, understand alternatives, and heed consequences. In the West the problems of freedom and the feminine are closely related, since mature freedom requires ending patriarchy's social, psychological, and religious repression of the feminine.

FREUD and Freudian psychology stand to Jung and Jungian psychology, Jungians believe, as Ptolemy and the geocentric theory stand to Copernicus and the heliocentric theory. Freud made ego the regulating center of the psyche. Jung saw ego as but one of many subordinate archetypes in a psyche whose ultimate arbiter is the self. Similarly, Jung saw in the oedipal complex one of many archetypal patterns of transformation, not the single dominant pattern Freud saw. Jung's psychology is revolutionary not only in respect to Freudianism but also in relation to most Western philosophies and orthodox theology, both of which make ego values supreme and pattern the Deity upon ego. What Freud called the superego, Jung identified with the Father archetype and collective thinking. Freud's id can partake of Jung's shadow, although the shadow is potentially a more creative and more conscious function than the id. Freud's theory of libido reduced all energy to sexual energy, whereas for Jung the sex drive was only one manifestation of psychic energy. Freud's psychology is oriented toward the masculine and men, particularly in their roles as fathers and sons. Consequently, it neglects the feminine and women, magnifies the oedipal complex, ignores the contrasexual anima and animus, and rejects the ideal of androgyny. The crucial absence of an overarching, regulating self archetype in Freud reinforces and expresses his essential atheism and his reductivism.

GODDESS OR GAIA represents the archetypal life force active in nature on planet Earth. The Goddess, Mother Nature, regulates the biosphere to keep various life forms in an overall symbiotic balance. The Greeks revered Gaia as the primordial Mother who nourishes and cares for all life. According to Hesiod's Theogony all the gods originated from her. Elements in Christian civilization continued to appreciate nature, though monotheism denied nature personification in a goddess. The devaluation of the goddess allowed the desecration of nature and environmental degradation that has spread across the planet in the wake of Western technology.

GODHEAD, for Jung, refers to the unfolding Quaternal totality of the Western divinity. It encompasses the Father, Son, and Holy Spirit or Paraclete along with a repressed Fourth figure whom Jung in different phases of his thought speculated must be either Lucifer or Mary. While Godhead's structure mirrors the four-part structure of psyche, its activity manifests the three stages of the dialectic of individuation: 1) The Father or the rude beginnings of divine individuation; 2) the polarized opposites of the dual archetype of the two hostile brothers, Christ and Lucifer; 3) and the consummation of the process in the Paraclete who epitomizes God the Artist and embodies the wiseperson archetype. The Paraclete as *anima* and Father-Son-Satan as *animus* form the yin and the yang of that divine androgyne and symbol of the self that is Godhead. The Western Godhead's efficacy in imaging and facilitating the self and its individuation accounts for its tremendous historical dynamism. As an image of the individuating self, Godhead finds its counterforce and collective shadow in the devouring uroboros.

Seen philosophically, a Godhead can represent the Universal Self or Supreme Identity or Actual God. Godhead technically refers to the essential nature, character, and structure of the deity or complex of divine figures that shape and dominate a culture or civilization. Godheads form through a coalescing of cultural traditions, feelings, intuitions, and ideas spurred by the activation of dynamic archetypes. The formation of Godheads mirrors a fundamental coalescing process in life and the universe wherein disparate parts gather and combine into a novel whole.

Just as each individual person forms his/her personal identity from a collective of voices, psychic functions, and archetypes coalescing within the self, advanced cultures form a divine identity or a Godhead from the collective of dominant archetypes coalescing within their culture. A civilizational Godhead is not usually tyrannized by single psychic archetypal function like the ego, as appears to be the case with Allah. More often we see a team of rival archetypes: consider, for example, the Hellenic and Hindu pantheons. The activity of the family of archetypes that form a civilizational Godhead determines the entire civilization's conceptions about Actual God.

GOD-IMAGE is present in the psyche of all human beings, although each individual and culture deals with it differently. Like the self which it reflects, it precedes and transcends individuated consciousness. Its universality in the human psyche explains the universality of religion in human cultures. It is not an image of Actual God, but an expression of the ordering principles of Actual God present in complex entities in the universe. Individual god-images are shaped by the culture context that provides the language, values and conceptual framework for dealing with them. The transcendent function which activates opposites becomes active when the god-image is energized. Losing sight of the god-image, Jung believed, is a major cause of the spiritual malaise of modern man.

HISTORICITY. Heidegger defined destiny and freedom in respect to an individual's temporal existence, or historicity, as distinct from an absolute time set by either the omnipotent, omniscient perfectly good God of Christian orthodoxy or the Newtonian mechanistic time. In Heidegger, time's basic tense is the future: reaching out (by teleology) from the future time turns back to assimilate the past which has created the present. One's destiny, one's personal timeframe is finite, it has a beginning and an end in death, and there is no immortal soul of course. The authentic individual must accept as his own the history or past that has been given him and chose to be responsible for it. That means we are to enact freely in our time the destinies that are given us by no choice of our own. Authenticity is the gift of decisive identity.

HOLY SPIRIT or Paraclete (literally, companion) is the neglected fourth member of the Godhead. It functions to resolve the dualistic conflicts of the two hostile brothers in a *coincidentia oppositorom* that consciously individuates the whole. Its guiding, healing, and motivating work incorporates that of the transcendent function, intuition, and eros. Being closely associated with Greek Sophia, the Old Testament wisdom figures, and the wisewoman and *anima* archetypes, and playing the roles of guide, comforter, and inspirer, it is predominantly feminine in function. The Paraclete acts as the single voiced *anima* or yin of the Christian God, who brings individuated wholeness to Godhead through *conjunctio* with the yang or multi-voiced animus-spirits of Father-Son Satan: thereby the Divine Androgyne is formed. As anima of God, the Paraclete manifests God as artist and the *lumen naturae,* which explains why Milton intuitively identified his muse with it. The archetypes constellated in *King Lear* form a Quaternal Godhead wherein Lear manifests the male archetypes and the Paraclete is manifest through Cordelia. Largely ignored by orthodoxy, the Paraclete personifies the neglected soul of Western culture and may provide the keys to its transformation and continued individuation. Jung speculated that the future of Christianity depends upon transcendence of the strife of hostile brothers (ego and shadow / Christ and Satan / Logos and Pathos / form and substance) through activation (as Eros and Sophia) of the Paraclete.

IDEOLOGY promotes a war of opposites with each side seeking the destruction of its opponent. By contrast, religious motivation accommodates essential opposites to transcend or integrate them in a process that unfolds the civilizational Godhead in time. Ideology tends toward the Manichean. With the decline of religion in the West and elsewhere, ideology seeks to fill the emotional niche left vacant by religion by giving people something to believe in and a sense of identity through rapport with a group. Ideology often arises from political ambitions spurred by economic desperation and opportunism, as is evident with Marxism, an ideology that demands religious fervor and commitment toward the ends of political power, economic control, and social regimentation. The initial Abrahamic religions, Judaism and Christianity, may be viewed as having generated two great Abrahamic ideologies. The first, Islam, is a religious ideology, the second Marxism is secular.

They are two of the four dominant ideological forces in the current political era. The third major ideology, Capitalism, which stresses the individual, where Islam and Marxism stress the group and define individuals through their membership in groups, owes its origins in part to the Christian emphasis on individual freedom and implicit commitment to the ideal of progress. The capitalists are correct in asserting that equality and freedom must include economic opportunity and freedom. However, when the economic freedom of the few trumps ethics, it diminishes freedoms for the many that are essential to maintaining a healthy, stable democracy. The fourth major ideology of modern times, positivism, arises in part from a rebellion against religion, and aggressive agnosticism about its claims. It strives to find values in scientific endeavor and method and substitute them for religious values.

IMAGES furnish a context and medium for archetypes. Jung discovered that archetypes and psyche itself operate imagistically through symbol, myth, and metaphor. Consequently, archetypes and the psyche are best approached and understood by imagination. For Jung, imagination is primary to and a precondition of thought and perception.

IMAGINATION is essential to all thinking beings, especially so for the artist. The artist's imagination can create out of pure feeling something tangible that can be sensed and heard physically and understood intuitively, if not always intellectually. It can transform what begins as in individual experience into universal art that many can apprehend and enjoy. Is the universe itself God's art making, an exercise of divine imagination within the medium of time? Just as the past is the realm of memory, the future is the realm of the imagination. Imagination is a unique and essential activity of consciousness. The soul is an image maker par excellence; it is the dynamism behind imagination. Imagination, of course, is not confined to artists and art. Scientists and businessmen need imagination, as does everyone in all walks of life. Because imagination is the work of the soul using material from the unconscious, it is essential to the individuation process where the wholeness we strive for is ultimately our unique personal work of art.

IMAGO DEI refers to cultural, rather than personal, expressions of the god-image. According to Jung, who followed Kant more than Plato, we cannot know God as metaphysical entity or *ousia*-that is, essential being. (Not a valid assumption in respect to The God in Time whose existence we experience directly.) Unto the unknowable entity we call God, Jung believed, we project archetypes that form our *imago dei.* From the self, the supreme archetype, we derive the idea of God; and the self provides our most adequate image or metaphor of God. However, other archetypes are projected unto the divine unknowable, for example the collective unconscious in elaborate pagan pantheons, or the paternal ego in Christian, Jewish, and Islamic monotheism. These polytheistic and monotheistic divine images form partial, and therefore inadequate, metaphors of the uniplural self. Jung saw the psychic imbalances, dysfunctions and diseases that afflict civilizations as symptoms of the inadequacy of their *imago dei,* or to use my term, the civilizational Godhead.

INDIVIDUATION is a dialectical movement from split and partial consciousness toward harmonized, fully differentiated consciousness that gives mature freedom and requires separation from the collective. It involves differentiation and then integration toward the goal of making the whole self conscious, or achieving wholeness. Individuation works against the counterforces of inflated ego and devouring uroboros. Spurred by the self, individuation proceeds by raising opposites from the unconscious and eventually conjoining them in order to manifest, give substance to, and finally harmonize the diverse parts of the psyche. In the individuation process ego becomes aware of (l) its separation from the unconscious uroboric state, (2) its function as the center of consciousness through identification with the hero archetype, (3) its need for substance or enrichment through opposition from the shadow, 4) its limited power through tragic experience, and finally, (5) through the work of *anima-animus,* aware of its interdependence with the whole self, thereby realizing the wiseperson archetype.

Individuation, moreover, can be viewed as a drama with the psyche its theatre and the self, as dramatist, orchestrating the archetypal players. Ego plays the protagonist, shadow the antagonist or comrade, *anima* the heroine and *animus* the champion. Like life and like drama, individuation is a dynamic process of continuous transformation and unfolding wholeness that would cease were it ever to permanently achieve a final, static goal. While the process varies among individuals, images of transformation, e.g., the journey, death, and rebirth, invariably signal its presence. An alternative view of individuation emphasizes not a transformation process directed toward wholeness but an inner deepening where *anima-animus* enables self to enter and enrich consciousness though epiphany, initiation, tragic sacrifice, and catharsis. In either case individuation by opening ego consciousness to self also opens it to God.

When we project self onto *imago dei,* we simultaneously project the individuation process. As a culture matures or individuates so does its *imago dei.* The climatic stage of divine individuation in Western culture is the incarnation and crucifixion; but its maturity lies in the future with the full manifestation of the Paraclete. In *Answer to Job*, Jung depicts the Judeo-Christian Godhead as an individuation process, set in mythic format, striving for a yet to be achieved wholeness, in short an unfolding God. The thesis of this work is that the fullest most authentic revelation of the divine individuation process is in the plays of Shakespeare, particularly *King Lear* and *The Tempest.*

INFLATION makes blocks individuation making wholeness impossible. The inflated ego aggressively expands its role within the psyche, often by identifying with an archetype or an archetypal function like the shadow. This identification alienates ego from other psychic functions such as *anima-animus*. Just as the self prompts individuation so the self's antipode, devouring uroboros, prompts inflation. The ultimate inflation, uroboric evil, occurs when ego denies all dependence on self and *unus mundus* to become a totally alienated, self-fed, self-consumed entity like a black hole. Uroboric evil makes the *imago dei* a likeness of a hypertrophied ego, thereby fashioning ego deities such as the *Summum Bonum.* In the less grandiose forms of inflation ego attempts to counter its fears of death and, more often, of failure in life. Inflation seeks security and certainty but actually generates alienation and obsession, which prepare the way for inflation's opposite, deflation or despair. Inflation can be seen as ego sinning against self, others, and cosmos. The remedy to the "sin" of inflation is not repentance but a humbling transformation of ego that opens it to the range of archetypes including shadow, *anima-animus*, and self, recognizes interdependence, and gives the faith or *pistis* to accept the uncertainties of life. In Renaissance literature alienation and inflation commonly reveal themselves in envy and malice. Inflation is central to the psychology of all comedy and tragedy. Comedy exposes and deflates ridiculous inflation. Tragic inflation may arise from flawed virtue in the hero or from inflated malice in the villain; tragic catharsis allows a purging of ego inflation to release the whole self.

INITIATION as a psychic quest enacts transformation and occurs when one breaks from the collective in order to set out on a journey to individuated consciousness. Instigated by the self, initiation begin with the death of a limited ego identity no longer adapted to life's challenges, and it ends with rebirth into a more comprehensive awareness. The initiate in formal rituals is accompanied by an initiator who mediates with the spirit realm-usually a shaman, priest, doctor, trickster, or saint-and who carries the projection of what the initiate will become. The psychoanalytic process is analogous to traditional ritualistic initiation, indeed Jung maintained that psychoanalysis is the only valid initiation process in the West today. Hermes is the god of initiation in alchemical and occult traditions. But advanced and primitive societies throughout the world have their own initiation rituals, including pseudo or shadow initiations into the collective. Initiation is a key process in Shakespeare, above all in *Hamlet* whose journey is an initiation where the protagonist frees himself from excessive attachment to the mother and moves toward individuation.

From *Henry IV* on initiation is a key element many of the greatest plays including *Midsummer Night's Dream*, *As You Like It, Twelfth Night, King Lear, Measure for Measure*, *The Winter's Tale*, and *The Tempest*. Except in *King Lear* where it comes at the price of the protagonists lives, initiation brings healing. Although Initiation is normally positive, but it can be a move away from consciousness into obsessive evil as when Iago seduces Othello in a retrograde initiation. In these cases it brings corruption, disease, and disaster.

INTERCONNECTION, TEMPORAL is closely related to and entails interdependence. Interdependence and interconnection sustain succession and continuum and make time the supreme relational process and universal regulatory system. A temporal, process universe is necessarily an interconnected, interdependent universe. The mystery of time is the mystery of interconnection and interdependence: how will events in the present affect the future, how do events in the present reflect the past? Interdependence and interconnection, (that which is interdependent is interconnected and conversely) are fundamental principles of Jungian thought on religion. They are even more fundamental for Whitehead and process philosophy, and for process theology as well which interprets love in terms of interdependence and respect rather than the dependence and obedience stressed by Pauline stasis theology. Jung's Quaternity symbolizes the interdependence and interconnection of the four archetypes that constitute the psychological reality of the Christian deity. Interdependence and interconnection are also fundamental to science: without interconnection nothing could be predicted because nothing would be related to anything else. The universal interconnection created by and characteristic of time makes tragedy, comedy, and religious meaning possible. Without interconnection there can be no tragic loss, comic gain, religious value, or scientific prediction. You cannot lose, gain, value, or predict anything with which you are not interconnected and interdependent.

INTERDEPENDENCE is a key to synchronicity and a fundamental assumption of analytic psychology. Jung stresses the psychic and physical interdependence of temporal events in synchronicity. Interdependence, however, is implicit in many Jungian conceptions. Wholeness can be viewed as the harmonious interdependence or homeostasis of parts within an organism, while individuation is the selfsame interdependence among the archetypes of the self. Likewise, androgyny realizes interdependence or syzygy of individuated male and female elements, *coincidentia oppositorom* realizes interdependence of opposites, and *unus mundus* realizes interdependence of physical and psychical. From a Christian perspective, interdependence requires and manifests love and combats alienating inflation. Interdependence is the ideal relationship of man to God and of ego to self as they work to co-create consciousness. Quaternity assumes the interdependence of the archetypes of Godhead. Like Quaternity, interdependence entails uniplurality and a philosophy of process. Moreover, it stands opposed to the alienation latent in uroboric monism or stasis, and active in radical pluralism or flux.

INTUITION is the psychic function that explores the liminal, determines inherent possibilities, grasps relationships and interdependence among parts, synthesizes opposites, and sees things whole. Intuition, through metaphor, finds new links between previously (and logically) unconnected ideas. Though Jung does not stress the synergy, intuition's reliance on metaphor and its power to connect parts into wholes makes it synergistic with imagination. Intuition, through imagination, generates visionary poetry. All vision, in the sense of prophetic vision, philosophic vision, or artistic vision, is largely a product of intuition. The inward vision manifest in Jungian psychology and associated with the Holy Spirit's wiseperson archetype relies on intuition. Identified with the god Hermes, intuition, a mental quicksilver, moves swiftly from idea to idea skipping logical connectives to leave the more ponderous thinking function far behind. As mystic channel to the realm of the archetypes, intuition is as essential to psychic and spiritual development as thinking is to intellectual development. Jungian psychology, in its effort to map the psyche, relies on intuition no less than on thinking.

INWARD VISION in its most basic sense, designates man's second sight of consciousness. Seen philosophically, it contrasts to Erich Neumann's old ethic of obedience and its opposite the ethic of rebellion. Inward vision realizes transformation by reconciling opposites, discerning new patterns, and linking parts to make wholes. The goals of inward vision are understanding in place of obedience, persuasion in place of command, mature freedom or individuated wholeness in place of arbitrary free will, and metastance (or overview) in place of inflexible commitment. Some of its associations are: The Holy Spirit, intuition, the wiseman or wisewoman archetype, *anima-animus*, *conjunctio oppositorum,* interdependence,

Kierkegaardean inwardness, Jung's analytic psychology, and process philosophy.

IRREVERSIBILITY is the essential character of time that makes time the creator and destroyer of order, value, and meaning as well as the only existential absolute. The irreversibility of time, not death or eternity, is the ultimate reality of the universe and our lives. Time's irreversibility, gives value and meaning to all things within time and the universe and thus acts as a metaphysical guarantor of the meaning of life and the validity of religious experience. Being the necessary condition for value and meaning, and therefore the absolute, it is identified with Actual God and makes time the essential power of Actual God. Growth toward consciousness involves acceptance of and creative interdependence with a power outside and greater than the ego and the self, *irreversible time. A*bove the quantum level everything and every event has its own singular place in space and in irreversible time which makes everything unique and in that way gives value to everything. Time as process means time is irreversible and some losses will be irreparable. Irreversible time and irreparable loss make tragedy possible, indeed inevitable. The crux of tragedy, its ultimate message, is ever the irreversibility of time and the indisputable fact of irreparable loss. Actual God, The God in Time, has made, by the irreversible nature of time, a universe where suffering, injustice, irreparable loss, tragedy and death are inevitable. The creative nature of time makes order, truth, value, beauty, and meaning not only possible, but evident to anyone. Moreover, time's irreversibility, that is a fixed, stable past, is essential for moral accountability as well as for the pursuit of truth, especially reliable prediction and scientific truth.

KING LEAR is, along with the Book of Job and the Gospels, one of three paradigmatic revelations of the archetypes of the Western Judeo-Christian Godhead. Lear goes through stages that manifest the Yahweh, Job and Prometheus archetypes, while Cordelia manifests the Paraclete and also Christ to create a more complete revelation than in any other mythic or literary source. Tragedy is the supreme literary art form of Western civilization whose development of tragedy far exceeds that in any other civilization. Tragedy's supremacy, and *King Lear*'s supremacy within the genre, derives from its revelations of God, the human condition, and, ultimately, its unflinching recognition of time's implacable irreversibility. The Promethean archetype provides the key to that recognition in *King Lear*, and the key to the Godhead archetypes that silently powered our civilization to become pre-eminent not only through science and technology but in the arts and philosophy. The Promethean archetype brings an unflinching recognition of truth which Edgar proclaims in the closing lines of *King Lea*r that serve as Lear's epitaph.

LOVE, or **EROS** is to *anima* what logos is to *animus* and pathos to shadow. Love moves us toward an ideal interdependence based on full awareness of self and other. Hence, love encourages dialogue, precludes projection and scapegoating, and curbs inflation and alienation. Love, creating interdependence, is the way *anima* confers value on others and the world. Though love works through *anima,* its ultimate source is the self's desire for harmony with the *unus mundus* or God.

LUCIFER is an aspect of the additional fourth member of the heterodox Jungian Quaternity. Manifesting the dark half of the dualist archetype, or the hostile brothers, Lucifer incorporates features of the rebel, the trickster, Dionysus, and the alchemical Mercurius. His creative and destructive sides are PROMETHEUS and SATAN. He stands dialectically opposite the Son-Logos or the divine ego confronting his obedience and ethereal spirituality with a shadow (or dark counter ego) of rebellious pathos and instinctive animal drives. Viewed externally he is God's scapegoat, but understood by Jungian psychology he is God's potentially creative shadow, which orthodoxy, through its perfectionist doctrine of *summum bonum,* insists upon repudiating and projecting on a putatively external scapegoat to deny the dark side of God.

MANDALA. Taking a geometric figure in with a circle squares or a square encircled with subdivisions divided by four or eight for its standard visual image, the mandala symbolizes the Quaternity or interdependent wholeness of the total psyche or the cosmos. In the West the Quaternal Godhead forms the supreme mandala.

MEANING in a generic sense refers the character and quality of a moment or series of events in time, their origins, intensity, relationships, consequences, duration, direction, interdependence with, but above all their uniqueness and singular place within the fluid processes that constitute the web of time. Meaning requires conscious reflection upon all the proceeding along with conflicts, and conflict resolution which exist only in time. Thus, meaning entails consciousness, and increases with consciousness. Time is the medium of order, meaning and value. Meaning, as in the phrase, "the meaning of life" refers to the possibility of realizing intents or values, establishing relationships, and fulfilling goals that are gratifying in a spiritual as well as emotional sense, and finally of connecting us with somethings higher: e.g. all humanity, all life or the Tao, the cosmos, Actual God.

MEMORY is our way of resisting time the destroyer, or retaining an aftertaste of our momentary experience of meaning, value, joy, pain, and beauty. In memory the past not only survives in the present, as Bergson noted, the past provides a framework that gives meaning and value to the present. The art of memory is essential to conscious thought, it engages the imagination, precedes and is more basic than logic or reasoning. A memory is more like an impressionistic painting than like a photograph. Memories are enlivened by consciousness through art. A memory is an artistic creation. Indeed, memory makes artists of us all.

MERCURIUS, the alchemists term for the god Hermes, is associated with the trickster, the shadow and the dark brother of the two hostile brothers archetype, as well as with transformation and the transcendent function. He has a fondness for sly jokes and pranks and plays the joker or the fool, as in King Lear's fool. He is an initiator and represents the most creative aspect of the brothers within the Quaternity.

METASTANCE is achieved when, having taken an initial stance that to the best of our knowledge seems valid, we stand back from our stance to gain a larger perspective that establishes a qualifying context. Metastance distances us from stance without denying its range of validity. Distance allows us to maintain tolerant skepticism, which counters dogmatism, despair, and therefore hubris. True metastance is characterized by a moral or, if you will, philosophical modesty wherein we recognize the need to ultimately rest every stance on a fideistic act. Metastance is Shakespeare's characteristic philosophic and artistic position. It is Shakespeare's corollary to Jung's *coincidentia oppos*itorum.

MYTHS in their primitive or natural form are stories containing archetypal encounters, patterns, symbols, and figures. Myths show archetypes unfiltered by consciousness and illustrate what can happen when they have free rein. By bringing the activity of archetypes into consciousness we are released from their compulsive hold the ego is allowed ego to tap their energies. We do not invent myths; we experience them as revelations from the psyche. Myths can be seen as metaphoric statements by the psyche about itself: God, like the psyche, reveals His and Her Self through myth. Consciousness can deepen and amplify myths and create new mythologems with which to restructure our perception of reality. Whether our myths are primitive or sophisticated, religious, political, philosophical, scientific, or artistic, we unavoidably perceive reality through the categories of myth.

NUMINOUS derives from numen or spirit and carries the meaning of "spiritual" without the baggage of its sentimental connotations. It can also imply holiness. Symbols are numinous, meaning that they are mysterious and cannot be precisely defined. Something is numinous if it originates in the psyche or unconscious, rather than the ego. Divine revelations are numinous. Jung called these phenomena dynamic effects not caused by will; they cannot be demanded; one must open oneself to them. He saw the experience of the numinous as the genesis of religion.

"Peak experience" is a loose contemporary psychological analogue to the numinous.

OEDIPUS COMPLEX, according to Freud, is the primary pattern shaping male ego development. In Jungian psychology it becomes secondary to the relationship with the mother; indeed, one of Jung's chief criticisms of Freud is that he recognized only one myth and one archetype: Oedipus. Jung saw the Oedipus archetype as shaping that stage of development where ego identifies with shadow to rebel against the Father or primal parent. Being central to the father and son or senex and puer relationships, Oedipus is also central to the relationship of God the Father and his two sons, Christ and Satan. The obedient Christ symbolizes repression of oedipal impulses and the rebellious Satan symbolizes living them out. Protestant Milton's inclination to subordinate the Son to the Father can be viewed as a theological extension of his strong repression of oedipal impulses. Freud's efforts to make the oedipal complex primary can be viewed as an unconscious attempt to repress Christianity and repudiate its theology: for Freud the oedipal complex shows the harmony of Father and Son to be impossible.

OLD ETHIC, as explained by Erich Neumann, relies on repression, obedience to the commands of authority, and adherence to orthodox dogma to maintain social order and personal sanity. Its controlling archetype is ego or superego, which appears first personified in God the Father and later becomes embodied in the Son as Logos. The old ethic protects ego in its immature development while simultaneously introducing a rigidity that divorces its values from their human purposes.

When discontented egos turn to the shadow for an alternative, the ethic of rebellion is born. The two hostile ethics, with their tools of contempt and envy, guilt and justice, perpetuate cycles of sado-masochistic manipulation that can be broken only when inward vision develops. The old ethic, the ethic of rebellion, and inward vision are linked to the fundamental philosophic stances on time: stasis, flux and process.

ORTHODOXY is the theological mainstream of Christian belief defined by canonical dogmas of the Roman Catholic Church, Eastern Orthodox churches, and most of Protestantism. At the core of orthodox dogma is the Nicene Creed, which holds Christ to be at once fully God and fully human, the doctrine of the Trinity, and the idea that God is the *Summum Bonum. Privatio boni* is an essential corollary to orthodox doctrine, even though Christians are not officially required to believe it. Orthodoxy posits a Trinity ruled by the Son-Logos and the Father archetypes where the Holy Spirit or Paraclete is marginalized. Ego, stasis, the old ethic, and the thinking function dominate orthodoxy and the Trinity dogma. Following logos, orthodoxy purges the feminine and intuition (wisdom) from the Holy Spirit and redefines it as a sexually neuter aspect of the Son and the Father. Jung sought to go beyond orthodoxy to Christianity's core myths and archetypes with his teaching that God is a Quaternity, not a Trinity. Jung's heterodox Quaternity is an attempt to restore balance and wholeness to Godhead so that it reflects the self-archetype and the entire psyche. Such a restoration, he hoped, could re-vitalize Christianity.

PARACLETE. Jung often used this term when referring to his heterodox conceptions of the Holy Spirit. Its simple literal meaning of companion eliminates the Trinitarian theological baggage freighted on the term "Holy Spirit" and more readily reflects the Quaternitarian feminine aspects of the Deity, particularly the Old Testament wisdom figure and the Greek Sophia. As companion, the Paraclete carries the archetype of *anima* and the function of divine soul. The Paraclete, Jung discovered, has this function in Hermetic and alchemical literature. It is also identified with Sophia or wisdom, and the wisewoman or wiseperson archetype. Truthful Cordelia manifests this archetype in Shakespeare.

PERSONA refers to the mask ego constructs and the roles it assumes to make specific impressions upon others, protect its privacy, and conceal its shadow. The darker the shadow, the brighter the *persona* must be to conceal it; and the weaker, or less individuated, the ego the stronger the persona. *Persona* and *anima* are contraries, the persona being concerned with the external and collective and the *anima* with the internal and individual. Being an archetype, *persona* is permanent feature of the psyche, like the ego, *anima*, and shadow. Yet it is the most protean archetype, it can alter greatly as it adapts to the vast array of human cultural and social conditions. Ego identification with persona leads to rigidity and loss of adaptiveness and humanity. This occurs often in Shakespeare's political and comic characters, for example Falstaff, Orsino, and Polonius, but also marks dark or tragic figures like Othello, Shylock, and Angelo in *Measure for Measure.* In orthodox theology the doctrines of *summum bonum, privatio boni,* and the Trinity form the *persona* of God. Insight and individuation require facing the activated archetypes *persona* conceals but which can lead us to the self and God.

PISTIS is a type of faith that contrasts to faith as blind submission to an external authority or faith as willed belief in doctrine. For Jung *pistis* refers to a conscious trust in the creative, sustaining powers of the self, or, to use his phrase, "fidelity to the law of one's own being." Seen in a specifically Christian context, it expresses the authentic love that comes not from ego but from self. *Pistis* requires authenticity and integrity.

PLEROMA is the fullness of being encompassing creation and nothingness. It incorporates a Gnostic notion of a oneness, a *unus mundus,* in which all opposites are subsumed. Mystical states involve perception of the *pleroma. Pleroma* also refers to the timeless realm of the archetypes. Jung's predilection for the term is one of many indications of the profound influence of Gnosticism on his thought. For Jung wholeness appears to involve the apperception of the *pleroma*.

PRIVATIO BONI, the doctrine that evil is merely deprivation of good, was promulgated by Augustine and others to explain evil without compromising the perfect goodness or absolute power of God. It assumes the identity of Being and the Good thereby making the Supreme Being the *Summum Bonum. Privatio boni* is one of two common solutions to the problem of evil; the others are (1) absolute dualism and (2) *coincidentia oppositorum,* which means that a single principle accommodates interdependent good and evil held in dynamic tension. An advocate of *coincidentia oppositorum,* Jung rejected *privatio boni* on the grounds that good becomes unreal when *privatio boni* deprives its opposite, evil, of substance. Jung's rejection of *privatio boni* links him to Freud and the existentialists through shared belief in the tragic and irrational dimension of life. Loath to recognize the causes and ramifications of that rejection, many Jungians dismiss or downplay it. They fail to see that Jung's rejection of *privatio boni* for *coincidentia oppositorom* results not from his idiosyncrasies but from his theory of consciousness: consciousness develops as opposites are constellated from the unconscious, differentiated into powerful and real real alternatives, and eventually (if the counterforce of uroboric unconsciousness is overcome) joined in a conscious *conjunctio oppositorom.*

PROCESS transcends the duality of stasis and flux through *conjunctio oppositorom.* Process assumes the universe is developmental and allows for it to be eschatological; it entails uniplurality, transformation, interdependence, homeostasis, and *coincidentia oppositorom.* It is associated with the Paraclete in the unfolding Quaternity, individuation, the *anima-animus,* and self archetypes, the intuitive function, synthetic thinking, inward vision, and the tragic outlook. Uroboric evil is for process the prime evil. The characteristic philosophical stance of Jung and of many great artists including Shakespeare is process; though Shakespeare appears to more fully engaged in process than anyone else. Process contrasts with and incorporates stasis and flux, the three align with the present, past, and future tenses. Philosophers tend to find their niches on the stasis, process, flux spectrum. Whitehead is the dean of process philosophers, although he did not designate this term for his philosophy, but called it the philosophy of organism. Life and all organisms are preeminently advanced processes, more complexly and fully so than inorganic structures like the stars and planets which are also processes.

PROJECTION in the psyche works rather like a slide projector and a screen; however, the screen is usually another person, and the image projected is associated with an archetype activated in the psyche. For example, a man will often project his shadow upon an enemy or his *anima* upon his wife or lover, seeing in another his own unintegrated negative or positive characteristics and complexes. Individuation requires withdrawal of projections so we can consciously integrate the archetypes and see other persons as they are. Projection can also occur between parts of the psyche as in the case of Lady Macbeth, where *animus* is projected onto shadow.

PROMETHEUS rebelled against Zeus to bring fire and other benefits to mankind. He was punished by being chained to a rock with a vulture tearing at his liver. Through defiance of the father (Zeus), Prometheus became a creative, revolutionary force and benefactor of mankind who sweeps away what is atrophied and outmoded. The Promethean archetype dominates the final stage of Lear's quest for consciousness through his confrontation with injustice, evil, death, and ultimately God. Prometheus' name means forethought, he is commonly identified with light and fire, and he has often symbolized consciousness.

Lear in his last scene exhibits Prometheus's insight into cosmic justice, his sense of divine betrayal, and his defiant assertion of the value and dignity of human life. But the Promethean archetype in *King Lear* derives its peculiar significance and impact from the largely unconscious, mythic way Shakespeare the artist relates it to the Job, Yahweh, and Christ archetypes to form a comprehensive vision of (the Judeo-Christian) God. The second brother of the Quaternal Godhead could appear as Prometheus. Prometheus, the fire bringer, personifies creative insight and destruction opening the door and clearing the way for progress. Prometheus represented a tectonic shift away from Satan, a wholly destructive face of the hostile brothers archetype, into to a creative or revolutionary form of destruction that clears a path for new good. In the post 1500 era the Promethean archetype became a powerful driver of technology and of political revolutions against the oppression of a stagnant status quo. Prometheus embodies the psychic dynamism that motivates men to seek freedom from all forms of slavery. It also motivates the quest for scientific knowledge and truth.

PSYCHE, in its restrictive meaning, refers to *anima* or soul. In the broader meaning Jung ordinarily employs, psyche encompasses all archetypes, psychic activities, processes, and possibilities both realized and unrealized, conscious and unconscious. It subsumes both the self and its uroboric counterforce. The ego and the conscious personality are only a small part of the total psyche. While ego identifies with one sex or the other, the psyche itself is always fundamentally androgynous. From the regulating, unifying center of the total psyche, the self, we form the idea of God. Psyche includes the self, but its larger than self insofar as psyche is impersonal and, for Jung, ultimately unknowable.

QUATERNITY is for Jung the psychological and archetypal reality of the unfolding Divine Self that the orthodox dogma of Trinity, working like a persona, conceals. It involves a heterodox four-part *imago dei,* or Godhead, reflecting the four-part totality of psyche and self. Jung sought in Quaternity a truly uniplural *imago dei* that incorporates the psychological, philosophical, and theological advantages of both monism and pluralism (or monotheism and polytheism) while avoiding the pitfalls of each. Quaternity gives wholeness to deity by restoring the feminine with the Paraclete-*anima*, the material and the instinctive with Satan-shadow whereby recognizing the potential for evil and irrationality. Early Christians had great difficulties with their Trinitarian Godhead because it offers a psychologically simplistic *imago dei* that presents the Persons of Godhead as similar to two male egos within one being and neuter third whose function is vague and problematic. Quaternity, by presenting the Persons as separate archetypal functions and stages (primal parents, ego, shadow, and *anima*) within a homeostatic Divine Self, can dispel the confusion, obscurity, and alienating incomprehensibility that surround the Trinity. Quaternity can bring order to the chaos of polytheism when we realize that the lesser gods or archetypes are subsumed under the four prime archetypes, parents, ego, shadow, and *anima*, of the unfolding four-part self.

SELF in its most general meaning refers to the essential wholeness, or unifying dynamism, of an organism. In its more restrictive sense, self refers to the psyche's superordinate, transconscious regulating function that activates archetypes, works toward the *coincidentia oppositorum* or psychic homeostasis, combats uroboros with its temptation to inflate, and thrusts ego down the path of individuation. The self in its wholeness is both an ultimate goal and the dynamism impelling us toward that goal. Self and *imago dei* are interdependent so that as awareness of self develops the image of God so that it more reflects the self. To put this another way, God, the self of the cosmos, interfaces with man through each person's self. In Jung's psychology the self is the energetic center of the psyche; self stands to ego as sun to Earth. This means that Jung follows a paradigm radically different from that of Freud (and most traditional Western thought), for whom ego is paramount. The paradigm shift from ego to self (Jung's Copernican revolution in psychology), along with its controversial implication of an androgynous rather than a single-sexed human psyche, accounts for much of the resistance Jungian psychology encounters.

SENSATION. Perceiving what is actual and physically substantive rather than possible or apparent, sensation is the psychic function opposite intuition. The most basic and primitive of the functions, sensation is the first aspect of consciousness to become differentiated. It is associated with the Father and the animus in Hebrew myth; in Greek myth it is most commonly linked to Gaia, Artemis, and other feminine nature deities.

SHADOW should not be taken literally, since a physical shadow is nothing but a deprivation of light. The term is a metaphor for a dynamic realm of the psyche cloaked in shadow, and for the archetype ruling that realm. The shadow contains energies that ego has difficulty accepting or developing, and therefore relegates to the personal unconscious. In contrast to the contrasexual *anima-animus*, shadow is the same sex as ego and includes qualities that could give strength and substance to ego. For example, the shadow often incorporates that Promethean-Luciferian protest and anger necessary for healthy reactions to injustice and intolerable situations. Sometimes, shadow contains the playfulness repressed by that overdevelopment of the ego function Freud called superego; some expressions are the adventurer, trickster, and prodigal son. Shadow has a largely ignored light side manifest in the sense of humor, and the mischievous creativity associated with Mercurius. Ego either engages shadow in a dialectic and, through mediation of *anim*a, transforms and integrates its protest and playful energies thereby moving toward individuated wholeness, or it combats and deliberately represses shadow.

Repressed shadow can break from its prison cell and take possession of ego, sometimes in an outright Satanic form. Ego, possessed by shadow, embraces an ethic of rebellion that turns the old ethic upside down by calling good whatever was formerly deemed evil. If shadow remains repressed, ego will project it upon those regarded as alien or enemies who then become scapegoats. Projected upon scapegoats, the shadow can be warred against. Projection of shadow explains most personal antipathies as well as most prejudice and persecution. For Jung the healthy way to deal with shadow is not repression but a creative transformation that gives substance to ego by rendering shadow more conscious.

In addition to the ego's personal shadow, the self has a counterforce, the devouring uroboros, which contains its collective shadow and impersonal, evil *anima-animus*. Unlike the personal shadow whose evil is operative, the uroboric counterforce holds no repressed elements that can be transformed into new, conscious good. Rather it is a regressive pull opposing the self's drive for individuated consciousness.

Because, resisting consciousness, it never submits to transformation, the uroboric counterforce is an absolute, though not a metaphysical, evil.

SIN, for Jung, violates the self by choosing lesser over greater, part over whole. Sin in a religious sense is an idolatry that defines God not as God is revealed to us, through the self, but through one of its subsidiary archetypes, most commonly the ego. Jung's views differ from traditional concepts where God is a projection not of self but of ego and sin is impurity or, most commonly, disobedience. Alienation and inflation, not impurity, are the prime symptoms of Jungian sin. Jung was more concerned with healing the alienation, inflation, and other psychic imbalances that come of putting part before whole than he was concerned with soteriology, let alone purification.

SON, Messiah, Christ, Logos is the second Person of the Godhead in his general form, during his earthly mission, and after crucifixion. Orthodox theology identifies the Son with the light of logos, which embodies the rational ego and the thinking function of deity. The Orthodox Son, Jung believed, is an inadequate symbol of wholeness since, sacrificing instinct to law and substance to form, it loses the passional and elemental side of the psyche associated with Lucifer, Mercurius, and Prometheus. Both halves of the brothers archetype, the Son and Lucifer, are needed for wholeness. By emphasizing the light half of the psyche and of the brothers archetype, Christians activated the destructiveness of the other half, which set up an unacknowledged dualism. The Job archetype in legend and literature, epitomizing perfect faith and obedience, inspires the orthodox Son.

SOPHIA AND WISEPERSON ARCHETYPES inform the third person of the orthodox trinity. It initially coalesced out of Hebraic and Gnostic feminine figures associated with Yahweh such as the Shekinah, which refers to the glory or light of the divine presence and, like the Holy Spirit, is often represented by a dove. Sophia is the goddess of wisdom, from which the term philosophy derives. She is found in the wisdom books of the Bible, and its apocrypha, and is associated with both the wise king Solomon and the feminine attributes of Yahweh. Sometimes Shekinah and Sophia are seen as companions to the deity. The ancient Jews and eastern Christians revered Sophia; the Byzantines named their paramount cathedral Hagia Sophia. But in the west she was eclipsed and displaced by Mary. Mary and Sophia represent different archetypal aspects of the feminine. Mary is the mother archetype while Sophia alternates between the wise daughter, as with Athena (or Cordelia), and the wisewoman. Both are *anima* figures but the wisewoman is a grandparent figure, correspondent to the wiseman. Sophia as Paraclete serves as the *anima* of God, just as the son serves as the ego and Satan or Lucifer is the shadow. Together with the father they are the functioning archetypes that shape the actual Quaternity of the western Godhead, as distinct from the Trinity which is merely a canonical construct.

SOUL and self act, for Jung, within the psyche as God acts in the cosmos or life on Earth. Both the self and the soul unify the whole person, the self being more a thinking-intuition unifier, the soul being more a unifier through emotion and sensation. However, feeling is not the entirety of soul, for soul involves the totality of our experience, our decisions, our character especially, and our very being. Soul includes feeling, thought, will, sensation, intuition, imagination, and the mystical sense or the *unio mystica*. It is the coalesced sum of our feeling life, and the essence of our very existence within and across time. Suffering, work, genuine achievement, artistic expression, solitude, and joy enliven and enrich the soul. The soul is an image maker par excellence, it is the dynamism behind imagination and the inspiration of artists. In the West there are two major conflicting conceptions of the soul. The two conceptions of the soul entail two different conceptions of time, stasis time grounded in eternity and time as process. The Platonic conception places the soul in eternity making it immortal; whereas in the Hebraic view the soul as developmental not immortal, it requires the process of time to emerge, grow, and mature.

SPIRAL is the third major image or paradigm for time. Time is viewed as cyclical by Plato, the Hindus, and many other ancient sources. Science tends to view time as linear and focuses on the arrow to time. Whitehead, and in some respects Jung, view time as a spiral. For Whitehead time as a creative spiral is a crucial concept. The spiral combines a line and a circle, being both yet something different and greater. The spiral is not fully predictable and so fosters creativity and allows for adventure.

SPIRIT is the active force, the energeia, associated with *animus* or the masculine aspect of soul. Spirit is the vital spark of life that sets the direction and speed of time by powering the flow of time in sentient beings. Jung saw spirit as the non-material, energetic aspect of a sentient being and links it with consciousness, purpose and the *animus*. Spirit and soul on one hand and body on the other impact time differently. The time of body and matter is the time of clocks, the calendar, and nuclear decay; it is steady, predictable and uncreative. The time of spirit, and of soul, is creative and unpredictable. It is the time of dreams, visions, and synchronicities where intuition and feeling can grasp connections not immediately apparent to rational thinking.

STASIS views the universe as fundamentally static, monistic, eternal, rational, and perfect. It is compatible with the Son as Logos, the ego, the thinking function, deductive thought, absolutism, monotheism, the ethic of obedience to authority, Christian orthodoxy, a comedic outlook, and the doctrines of *summum bonum* and *privatio boni.* A one-sided philosophic commitment to stasis tends to force its opposite, flux, into shadow where it undermines stasis with covert dualism.

SUBSTANTIATION refers to the alchemically related process of actualizing good through a *coincidentia oppositorom* with the operative evils of shadow and suffering. Without shadow and suffering to substantiate good (and the ego) by building moral substance, the good remains an empty ideal or an unfulfilled potential. Moreover, unless substantiation anchors ego in reality, ego falls prey to uroboros and drifts into alienated inflation. Substantiation is, therefore, a basic feature of individuation. The alchemical term for substantiation is *coagulatio.* As shadow and suffering substantiate ego, so encounters with profound suffering and death substantiate the soul.

SUMMUM BONUM or supreme good is a conception of divinity stemming from Platonism and adopted by the Church fathers who sought to establish the moral perfection of the Christian deity. The belief that God is the omnipotent *Summum Bonum* makes it logically necessary to reduce evil to *privatio boni.* Jung condemned the doctrine of *summum bonum* as a projection of inflated, hypertrophied ego upon the *imago dei.* For Jung God is the *Summum Coincidentia Oppositorom* that encompasses everything and its opposite in a single interdependent reality and strives not for one-sided goodness but for plenary consciousness. Jung blamed the *summum bonum* doctrine for setting impossible standards of perfection that induce guilt and despair and thereby incapacitate us for individuation which requires dealing creatively with real evils. The doctrine lies behind the popular and deleterious expression, *"omne bonum a Deo, omne malum ab nomine"* (all good is from God, all evil is from man). Therein God gets credit for everything good, and man, with no stable good of his own, takes the blame for evil. Jung's rejection of the *summum bonum* doctrine decidedly places him outside orthodoxy and aligns him with the Gnostic tradition.

SYMBOLS express themselves as analogies, often appear as images, have intents that can be difficult to peg down, and they manifest the archetypal structures and energies of the unconscious whence they spontaneously arise. Unlike elements of consciousness, they relate to the whole psychic system, conscious and unconscious, not just the conscious ego. Multifaceted interpretations are a mark of true symbols. Being essentially ambiguous, they elucidate something unknown by more or less apt analogies in order to compensate for and ultimately transform conscious one-sidedness with a *coincidentia oppositorom.* Since their function often is to compensate for the inadequacy of rational concepts, once they are fully translated into rational terms they cease to be symbols and become mere signs. Symbols are most effectively approached through amplification of the images wherein they are embedded.

SYNCHRONICITY is an acausal connecting principle posited by Jung to reveal and explain meaningful connections bridging the psychic and physical worlds. *Synchronicity may be defined as: simultaneity fraught with psychological meaning that discloses significant qualities of a moment or period of time.* Synchronicities manifest the interaction and/or parallelism of the subjective with the objective. Jung's conceptions of synchronicity are, I believe, best approached as attempts to achieve a deeper, more comprehensive understanding of: 1) the complexity of time and causation; 2) the interrelationships of mind and matter; 3) the unique character of moments in time; 4) how constellated archetypes of the *unus mundus*, can shape and cause events. Synchronistic thinking reaches across the structures of mechanistic time and causation to show us that constellated archetypes can give us unusual experiences that shed meaningful light on significant challenges in our lives. These meanings yield novel insights that can become gifts of the self to the conscious ego for the purpose of opening opportunities to further individuation. When we notice synchronicities, it is likely because an active archetype pre-consciously organizes our train of thought. Jung believed in a religious implication of synchronicity, that it offers evidence of an underlying interconnectedness of events in the psyche and in the physical world. It also exhibits the universe's duality of macrocosm and microcosm, mind and matter.

SYZYGY is a Gnostic term for the divine androgyne, and for the pairings of each creative emanation of divinity Jung often appropriated it to refer to the *coincidentia oppositorom* of *anima* and *animus* or soul and spirit. Syzygy consciousness, which expresses the self and views everything in interdependent pairs or wholes, contrasts with one-sided ego consciousness, which denies interdependence in favor of dominance and dependence among alienated parts.

TARDIS designates a specific temporal context forming a discrete continuum of time, or a timeframe in the sense of a span of time that has its own intrinsic qualities, character, and rules. A tardis is like a magnetic or gravitational field, only operating within time and activated by archetypal forces and historic circumstances. Unlike an *umwelt,* a tardis is a phenomenon independent of the individual agent. It is, however, an interdependent, distinct section or field of the phenomenal time universe. Nonetheless, each tardis is unique because it has its own limits, a beginning and end even though both may be shadowy boundaries that shade into adjoining tardises just as gravitational fields phase into each other. Physicists have discovered that space is not empty, every area has specific activities and characteristics; similarly, time is not an empty medium but has its own metaphoric regions and geography. Tardises are temporal analogues to Whitehead's concept of spatial "societies" or distinct orderings of spatial perceptions. In terms of archetypal psychology, tardises occur when highly activated archetypes coalesce or configure to create a time field. Powerful archetypes can warp human existential time according to Jung, even as massive objects in space warp space-time according to Einstein. Seen another way, the coalescing of archetypes might be a psychic version of what happens with organic chemicals. The archetypes are like atoms of various elements which bond chemically to form complexly structured organic compounds. Identifiable time fields or tardises, possess independent, discernible characteristics, and are very roughly analogous to chemical compounds as well as gravitational and magnetic fields. The qualities of the tardis will influence and often profoundly shape the *weltanschauung*, or worldview, of those living within it.

THINKING FUNCTION seeks to comprehend and judge the world by means of logical inferences and rational analysis. It is associated with ego, *animus*, the ethic of obedience to authority, orthodoxy, the gods Zeus and Apollo, stasis, the logos, and the Son. The thinking function tends to place time in a stasis universe, which leads it to conclude that time is illusory. This contrasts to the sensation function where time appears to be a flux, and to the intuitive function where it appears to be a process. To the feeling function, the realm of the soul, time is the medium of meaning and value.

TIME in this work is seen as the absolute primordial regulatory principle, source of order, and the context of value and meaning; it is the field, structure and, with change, the dynamism of reality. Time is our indispensable experience, the medium of all relationships, and the sustainer of our existence. All life contends with time. Our experience in life is the multilayered and sometimes mysteriously interactive and interdependent product of the mind's struggle to make sense of and discover meaning and value within time. The three tenses of time structure and reflect our cognitive activity, especially our planning, hopes, and fears.

Time is the ultimate reality, the over-riding existential challenge. The one thing we cannot avoid, or evade and must ever contend with is Time, and that makes Time the essential function of Actual God. The spiraling creative dynamism of time and the medium in which Actual God acts are two sides of the same coin. Time is the means whereby God maintains order generates value by creating uniqueness which requires emergence, duration, change, and destruction. Nothing outside of time could have value. Time is the great value creator—and destroyer.

Time has fields like gravity. It creates interfacing regions in which a particular conditions prevail, those conditions being continuous interrelated, interdependent duration and change. Time itself is the hyperfield in which all change, duration, and indeed all that was, is, or will be exists and subsists. Philosophers, theologians, and scientists have argued that time itself does not exist or is unreal. Certainly, time is not a thing among other things. But rather than being unreal, it partakes of, regulates, and defines all reality, since all things are defined and differentiated in relation to and by their positions within time, that is within their temporal fields. Like reality and consciousness itself, time is essentially relational; time is the great relator of all things past, passing, and to come. Time is also the supreme differentiator since relationships differentiate all things and events.

TIMEFRAMES. Humans (as well as other intelligent beings, one supposes) perceive events within the contexts of timeframes, a generic term that includes but is not identical with tardises. Tardises refer to discernible structures, force fields around archetypes, or distinct ecologies, within the continuum of time that are independent of the particular observer, whereas timeframes are just the observer's ways of consciously organizing time. Like tardises, timeframes often phase into each other and overlap.

TRAGEDY has unique importance in the West, as is evident by our deeming it the highest form of drama and the apex of our literature. Tragedy stands apart as a cultural qualification and challenge to religious orthodoxy, and to stasis philosophy's claims for an omnipotent, omniscient *summum bonum.* Since tragedy is intrinsic to the human condition, the concept and experience are universally understood even though tragedy is an essentially Western art form. Tragedy focuses on and arises from challenges created by a disastrous break in the expected flow of time. Tragedy gives meaning to time's sharp breaks, harsh blows and irreparable losses. It helps redeem our humanity by allowing us to transcend suffering and evil through the hero's struggle to create integrity, truth, and beauty. Such is tragedy's value and lesson, a lesson that makes it the supreme expression of the spiritual genius of Western civilization and the power of the archetypes of its Godhead.

Tragedy, particularly Shakespearean tragedy, presents spiritual and moral struggles, questions, and lessons of vast consequence. It reminds us that evil can win and the triumph of good is ever uncertain. It reminds us we live in a universe where our decisions can bring irreparable losses. It reminds us we are responsible for who we are, even if not for what our lives become, because who we are depends on how we handle what life offers and thrusts upon us. It tells us that time's fateful irreversibility, not death or eternity, is the ultimate reality of our lives. But tragedy also means we will have opportunities to create value for ourselves and meaning for our lives through authentic responses to the challenges of irreversible time. Tragedy entails *decisive identity*, that our decisions and choices determine who we are.

TRANSCENDENT FUNCTION. The use of the term "function" is problematic since it creates a confusion with the thinking, feeling, sensation, and intuitive functions. Most often operating through the *anima*, the transcendent function works to achieve transformation and psychic homeostasis by activating elements from the unconscious or by casting up unifying symbols that enable the ego to transcend destructive, one-sided attachments as compensatory opposites enter consciousness. The transcendent function thus allows opposites to dialogue with each other and mediates them to overcome their dualism in a *coincidentia oppositorum.* The Paraclete, providing inward vision and self-tuned conscience, constitutes the religious manifestation of the transcendent function. Whether called Paraclete, *anima,* or transcendent function, its objectives, psychic depth and individuated wholeness, are opposed to the one-sided perfectionism sought by the old ethic of obedience to traditional authority and belief in orthodox dogmas.

TRANSFORMATION is a developmental process wherein opposites are constellated in archetypal images and symbols and then consciously united, thus unfolding self to ego. The symbolism of transformation is found in initiation rites, alchemy, myth, nuptials, and religious ritual. Transformation begins with an inflated ego alienated from self and shadow and proceeds through symbolic ego-death, or deflation, to an *anima-animus*-mediated rebirth into a new consciousness that incorporates creative and substantial elements from the shadow. In transformation we confront, suffering, evil, and darkness. The ultimate transformation and final darkness is death; through confronting death we strengthen and deepen our souls. Man has an instinctive drive toward transformations that realize wholeness and depth: Jung calls this drive individuation. Transformation of evil and shadow into a conscious good is the constructive way to deal with their darkness, the destructive way being to deny shadow and project it upon scapegoats. Transformation, rather than perfection, is for Jung the proper objective of morality and goal of the self and divinity.

TRICKSTER archetype appears in rebellious hero and anti-hero figures like Satan and Prometheus, or in playful shape-changers like Mercurius, and in prodigal son figures. Embodying the shadow, it is often associated with the dark member of the two hostile brothers who champions individuality against the collectivity of the father archetype, and thereby compensates imbalanced conscious attitudes. By upsetting established order, the trickster releases creative energies along with chaos and destruction. The trickster symbolically manifests phallos. The character Autolycus in Shakespeare's Winter's Tale is a perfect example of a trickster though Falstaff, Lear's Fool, and Hamlet display clear trickster elements.

TRINITY, the orthodox term for the Christian Godhead, represents, according to Jung, the *persona* of the Godhead, rather than the divine self, as dogma contends. The actual Godhead, or the divine self, must be a Quaternity in order to allow for both the feminine, *anima*, and the shadow. Trinity results from the inability to comprehend and face the full reality of the Godhead.

TWINS as TWO HOSTILE BROTHERS or as FRIENDS (feminine pairs are the harlot and the virgin, and the mistress and the wife) embody dark and light symbols of same sexed psychic opposites. These pairs exemplify the interdependence of opposites necessary to wholeness and the tension of opposites essential to individuation. Ego and shadow, thinking and feeling, are the two sets of opposites most often associated with the rival brothers, whose signal Biblical exemplars are Abel and Cain, Christ and Satan. Just as the Great Mother has a nurturing Madonna side and a devouring uroboric side, the Twins archetype can come as either the supportive friends or the combative enemies.

Christianity emphasizes antagonism and war between the brothers, while Greek paganism emphasizes cooperation, dialogue, and friendly rivalry. One result is two very different attitudes toward the ego-shadow relationship and toward homosexuality. The ego that represses its shadow inevitably projects and scapegoats it on others of the same sex, fostering distrust of same sex bonds. The ego that dialogues with shadow, seeks its cooperation, and learns to use its energy, will put a higher value on same sex bonds than the ego that is antagonistic to shadow. In the religious sphere the orthodox Christian emphasis on the brothers' antagonism creates an implicit dualism between Christ and Satan. The hostile brothers archetype has energized a centuries long antagonism between Judeo-Christianity and Islam. At the risk of transgressing PC, one could speculate that the archetype's two greatest historical exemplars are Jesus and Mohammad.

UNCONSCIOUS is a general term for those mental contents not readily accessible to the ego, it also refers to a vast psychic domain with a nature and laws different from the conscious domain and divided into a collective, or inherited, archetypal sphere and a personal sphere. Both the collective and the personal unconscious are independent of ego. The shadow connects ego with the personal unconscious, and *anima* connects it with the collective unconscious. For Jung God does not correspond with the unconscious, as in polytheism, so much as with the psyche's unifying center, the archetype of the self.

UNIO MYSTICA, a Western alchemical analogue to the *Tao, samadhi,* and *satori,* which in Jungian psychology represent that supreme form of integration in which the ego is subsumed in a self illuminated with consciousness. It can also refer to a union of self, cosmos, and God. If such a state can be attained, it would put us in direct contact with Actual God. Whether and to what degree this ideal is attainable remains problematic.

UNIVERSE is everything subject to time, and time is its ultimate principle and reality. The God in Time is the Actual God of the universe. There may be sub universes, but this universe is the only one we have and can know, and its supreme law is time. Time orders and unifies the universe and thereby makes it a universe.

UNUS MUNDUS, an alchemical term, denominates the inherent unity or interconnectedness of the world and the ground of all empirical being (Brahman) sometimes symbolized by the *anthropos* or cosmic man and finding a contemporary expression in the idea of Gaia. Known intuitively rather than through reason or sensation, it provides a metaphoric background for acausal theories of meaning such as synchronicity, implicate order, teleology, and mysticism. For Jung, healing, or restoring wholeness to the psyche, involves establishing harmonious interdependence with the *unus mundus.* Wholeness brings knowledge of God whom we encounter first in the self and who ultimately unfolds in the *unus mundus.* It entails the interconnectedness and interdependence of all strata of being, particularly body and psyche, matter and mind.

UROBOROS takes for its image a single snake eating its tail, which contrasts with the interdependent energies of the yin/yang or tao, imaged as two snakes eating each other's tails. In its most primitive form the uroboros is simply the "Great Round" where all elements, male and female, good and bad, and promotive of and inimical to consciousness are mingled. As such, uroboros is an encompassing deity of the polytheistic pantheon. Epitomizing total self-sufficiency, uroboros images the archetype from which Plato derived the notion of *ousia* or being.

The self constellates opposites to create consciousness and thus spur ego to release from uroboros, to advance individuation, and, ultimately, to achieve conscious interdependence with the *unus mundus.* However, when uroboros exerts a regressive pull after ego's emergence, it becomes the devourer (or dragon) archetype that subverts consciousness and interdependence while promoting alienation and inflation. Moreover, devouring uroboros, functioning like a Freudian death wish, incites those destructive autonomous complexes, or psychoses, which find their definitive symbol in the unbearable image of Medusa.

It is the devourer who, under the inflated guise of a quest for perfection and purity, tempts the ego of man and man's gods to withdraw from interdependence with the *unus mundus* into alienated self-fed, self-consumed uroboric evil. In this role it becomes the evil *anima*, collective shadow and archetypal counterforce (sometimes symbolized by Leviathan) to Godhead and self. Although the self seeks conscious union of its inner opposites, union with the self's archetypal counterforce negates consciousness and the self's essential drive for individuation. Thus, the devouring uroboros is an implacable enemy to the self, and its evil is, unlike the operative evil of the personal shadow, absolute. Absolutely evil though it may be, uroboros is neither a useless privation nor a dualistic opponent to the good. It is not a metaphysical evil, a notion rejected by Jung and Whitehead. Overcoming the devourer's archetypal counterforce proves indispensable to the self's (and Godhead's) task of individuating consciousness.

VALUE requires feeling, feelings are ever fleeting and essentially temporal. Time and feeling are the form and substance of value, there is no value without time shaping feeling and feeling giving substance to time. Nothing outside of time could have value. Time is the great value creator—and destroyer. Value creation and destruction are the selfsame temporal process. Things that have value are valuable only because they can be, will be, and are transformed or destroyed: by their transformation and/or destruction they create novelty and scarcity, two prevalent measures of value. Time, the *creative continua*, ceasely creates novelty and with it new value. Duration sustains the value time creates.

The uniqueness of things is an essential aspect of their value, and another way that The God in Time, creates value. Time's irreversibility, as Augustine well understood, makes everything unique and gives unique value, or novelty, to every moment. The God in Time, acting through time, creates all value; as Whitehead indicated, creativity is this God's defining activity and character. A universe without time would be without value. There would be no differentiation between any single part and another, and no consciousness or separate centers of consciousness, e.g. minds, to distinguish and evaluate anything. Value, good, evil etc., depend on temporality and cannot exist without it. Those who argue that "God exists" are likely arguing that we live in a universe where value exists.

WHOLENESS AND INDIVIDUATED WHOLENESS, like mature freedom, denote a state in which ego, through *anima-animus* mediated integration of shadow, incorporates the whole psyche in its balanced, harmonious, fully differentiated, and essentially androgynous consciousness which brings psychic health. Individuated wholeness does not eliminate conflicting opposites, for tension is essential to life; it encompasses them in an interdependent, homeostatic system symbolized in alchemy by the *hierosgamos.* Because it requires interdependence, individuated wholeness is better understood as uniplurality than as simple unity. The ripening of the individuation process, individuated wholeness stands immediately below *unio mystica* as the highest state possible for man. Like *unio mystica,* its value as a goal for the individuation process resides in our inability to realize it permanently. In a generic sense, wholeness consists of the integral relationships between an individual's parts that is greater than and different from any of the parts or the sum of all the parts. The wiseperson archetype, dramatized in Shakespeare's Prospero, embodies wholeness. The orthodox Christian Godhead troubled Jung because it excluded two elements essential to wholeness, the feminine and the shadow or evil. He opted for a Quaternity over the Trinity to achieve wholeness in the Godhead. Shakespeare's vision of Godhead in *King Lear* achieves wholeness by relegating incorrigible evil to Lear's two wicked daughters, incorporating the positive feminine in Cordelia, and giving God a Promethean shadow in Lear's final defiance.

BIBLIOGRAPHY

Augustine, St., *Confessions,* (Book 11), Trans Maria Boulding, Hyde Park, NY: New City Press (2002).

Aziz, Robert, *C. G. Jung's Psychology of Religion and Synchronicity*, Albany, State Univ of New York: (1990)

Beebee, John, *Integrity in Depth*, New York: Fromm, (1992)

Bergson, Henri, *Creative Evolution*, New York: Dover (1998)

Time and Free Will, New York: Martino Press, (2015)

Bloom, Harold, *The Western Canon*, New York: Harcourt Brace, (1990)

Shakespeare: *The Invention of the Human*, New York: Riverhead, (1998)

Bohm, David, Wholeness and the Implicate Order, London: Routledge, (1980)

Bradley, A. C., *Shakespearean Tragedy*, London: Macmillan, (1904)

Susan Cady, Marion Ronan, Hal Taussig, *The Future of Feminist Spirituality*, San Francisco: Harper & Row, (1986)

Campbell, Joseph, *Man and Time*, New York: Bollingen Foundation, (1957).

Corbett, Lionel, *Psyche and the Sacred*, New Orleans: Spring Journal, (2012)

Dourley, John P., *The Psyche as Sacrament: A Comparative Study C.G. Jung and Paul Tillich*, Toronto: Inner City Books, (1981)

Driscoll, James, *Identity in Shakespearean Drama*, New Brunswick: Associated University Presses, (1983

The Unfolding God of Jung and Milton, Lexington: University of Kentucky Press, (1993)

Durkheim, Emile, *The Elementary Forms of Religious Life,* New York: Free Press, (1995)

Dyer, Donald R., *Jung's Thoughts on God*, York Beach, Maine: Nicolas Hayes, (2000)

Edinger, Edward, *Ego and Archetype*, New York: G. P. Putnam's Sons,

(1972)

The Christian Archetype, Toronto: Inner City Books (1987)

The Creation of Consciousness, Toronto: Inner City Books (1995)

The New God Image, Wilmette, Ill.: Chiron Publications, (1996)

Transformation of the God-Image: An Elucidation of Jung's Answer to Job, Toronto: Inner City Books, (1972)

Elton, William R. *King Lear and the Gods*, San Marino, Ca.: Huntington Library, (1966).

Frank, Erick, *Philosophical Understanding of Religious Truth*, Oxford: Oxford Univ. Press, (1959)

Fraser, J.T. *The Voices of Time*, New York: George Braziller, (1966)

Of Time Passion and Knowledge: New York: George Braziller, (1975)

Frey-Rohn, Liliane, *Friedrich Nietzsche: A Psychological Approach*, Einseideln, Switz: Daimon Verlag, (1988)

Frye, Northrup, *Anatomy of Criticism*, New York: Atheneum, (1966)

Fools of Time, Toronto: Univ. of Toronto Press, (1996)

Northrup Frye on Shakespeare, New Haven: Yale, (1988

Glatzer, Nathum, *The Dimensions of Job*, New York: Schocken Books. (1968)

Griffin, David, *Archetypal Process*, Evanston: Northwestern University Press, (1989)

Harding, Esther, *The "I" and the "Not I."* Princeton: Princeton Univ.Press, (1965)

Hawking, Stephen, *A Brief History of Time*, New York: Bantam, (1998)

Heidegger, Martin, *Being and Time*, San Francisco: Harper. (1962)

Identity and Difference, New York: Philosophical Library, (1960)

Heisig,

James, *Imago Dei, A Study of C.G. Jung's Psychology of Religion*, New Brunswick: Associated University Presses, (1979)

Helm, Bertrand P., *Time and Reality in American Philosophy*, Amherst: Univ. of Massachusetts Press, (1983)

Henn, T.R., *The Harvest of Tragedy*, New York: Barnes and Noble, (1966)

Hubler, Edward, *The Sense of Shakespeare's Sonnets*, Princeton: Princeton Univ. Press, (1952)

James, William, *Varieties of Religious Experience,* New York: Penguin, (1982).

The Will to Believe, New York: Dover, (1956)

Jorgensen, Paul A. Lear's Self Discovery, Berkeley: Univ. of California Press, (1967)

Jung, Carl Gustave, *The Collected Works (CW)* Princeton: Princeton Univ. Press, various dates, 20 vols.

Symbols in Transformation. CW 5 (1967)
Psychological Types. CW 6 (1966)
Two Essays in Analytic Psychology. CW 7 (1966)
The Structure and Dynamics of the Psyche. CW 8 (1969
The Archetypes of the Collective Unconscious. CW 9 Part I (1968)
Aion. CW 9 Part II (1968
Civilization in Transition CW. 10. (1970)
Psychology of Religion: East and West. CW 11 (1969)
Psychology and Alchemy. CW 12 (1969)
Alchemical Studies. CW 13 (1968)
Mysterium Coniunctionis. CW 15 (1966)
The Red Book. New York, W.W. Norton (1969)

Kerenyi, Karl. *Prometheus.* Princeton: Princeton Univ, Press (1963)

Kaufmann, Walter, *From Shakespeare to Existentialism*, Garden City: Doubleday, (1959).

Le Poidevin, Robin and Murray MacBeath, *The Philosophy of Time*, Oxford: Oxford Univ. Press (1993)

Mack, Maynard, *King Lear in Our Time,* New Haven: Yale Press, (1965)

McGinn, Colin, *Shakespeare's Philosophy*, New York: Harper Collins (2006)

Merleau-Ponty, Maurice, *The Phenomenology of Perception*, London: Routledge, (2013)

Mitchell, Stephen, *The Book of Job*, San Francisco: North Point Press, (1987)

Neumann, Erich. *Art and the Creative Unconscious,* Princeton: Princeton Univ. Press, (1959)
Depth Psychology and the New Ethic, Boston:Shambhala, (1990)
The Great Mother. Princeton, (1972)
The Origins and History of Consciousness Princeton (1970).

Nietzsche, Friedrich, Thus Spoke Zarathrustra, Walter Kaufman Trans. New York: Viking, (1968)

Pagels, Elaine, The Origins of Satan, New York: Vintage, (1996)

Priestley, J.B. *Man and Time*, Garden City, NY: Doubleday, (1964)

Price, Lucien, *Dialogues of Alfred North Whitehead.* Boston: Little, Brown & Co. (1954)

Samuels, Andrew. *A Critical Dictionary of Jungian Analysis*. New York: Routledge, (1986).

The Plural Psyche, New York: Routledge, (1989)

Sharp, Daryl. *C. G. Jung Lexicon*. Toronto: Inner City Books (1991)

Shakespeare, William, Collected Works, no edition specified. Works of particular relevance to the theme of time:

Antony & Cleopatra,
As You Like It,
Hamlet,
Henry IV, Parts I &II
King Lear,
Macbeth,
Othello,
Sonnets,
The Tempest,
The Winter's Tale,
Twelfth Night

Sherburne, Donald, *A Key to Whitehead's Process and Reality*, Chicago: Univ. of Chicago Press, (1966)

Stark, Rodney, *Discovering God*, New York: Harper One, (2007)

The Victory of Reason, New York: Random, (2005)

Stevens, Anthony, *Archetypes: A Natural History of the Self*, New York: Quill, (1983)

Teilhard de Chardin, Pierre, *Writings Selected*, New York: Orbis Books, (1999).

Todd, Peter B., *The Individuation of God*, Willemette Ill.: Chiron (2012).

Turner, Fredrick, *Shakespeare and the Nature of Time*, Oxford: Clarendon Press, (1971).

Von Franz, Marie-Louise, *The Shadow and Evil in Fairy Tales*. Zurich, Spring Publications, 1972

Individuation in Fairie Tales, Zurick:Spring (1977)

Interpretation of Fairie Tales, Zurich:Spring (1973)

Psyche and Matter, Boston: Shambhala, (1988)

Number and Time, Evanston: Northwestern Univ. Press, (1974)

On Divination and Sychronicity, Toronto: Inner City Books, (1980)

Shadow and Evil in Fairie Tales: Spring (1974)

Whitehead, Alfred North, *Process and Reality*, New York: Harper (1957)

Adventures in Ideas, New York: Macmillan, (1967)

Religion in the Making, New York: Fordham, (2001)

Science and the Modern World, New York: Free Press, (1997)

The Concept of Nature, Cambridge: Cambridge Univ. Press, (2015)

INDEX

www.ingramcontent.com/pod-product-compliance
Lightning Source LLC
LaVergne TN
LVHW070907240926
848811LV00017B/819
* 9 7 8 1 6 8 0 5 3 4 8 1 8 *